Warm and Cool Homes

Building a healthy, comfy, net-zero home you'll want to live in forever

To Stephen,
I hope you enjoy!
Wes Golomb

Wes Golomb

Copyright © 2021 Wes Golomb

www.TheEnergyGeek.org

All rights reserved. No part of this book may be used or reproduced by any means, graphic, electronic, or mechanical, including photocopying, recording, taping or by any information storage retrieval system without the written permission of the publisher except in the case of brief quotations embodied in critical articles and reviews.

Because of the dynamic nature of the Internet, any web addresses or links contained in this book may have changed since publication and may no longer be valid. The views expressed in this work are solely those of the author and do not necessarily reflect the views of the publisher, and the publisher hereby disclaims any responsibility for them.

Cover Photo is of Rob and Kerry Hunter's home in Canterbury NH. The house was built by RH Irving Homebuilders. Photograph taken by Wes Golomb

Prepared for publication by www.40DayPublishing.com

Cover design by Kayla Cloonan

Printed in the United States of America

DEDICATION

To my grandmother, Rose Golomb, my aunt and uncle, Sarai and Sam Zitter, and my cousin and friend Lee Paseltiner. You encouraged me to think for myself and to leave the woodpile a little bigger than I found it. I miss you and wish you were here to read this.

ACKNOWLEDGMENTS

Thank you to the many people who helped with this book. When I think of those who supported me in this endeavor, I realize how fortunate I am to have such a wonderful group of caring friends and family members willing to help me.

Specifically, thank you to:

Bob "The Builder" Irving for your friendship and for spending literally days with me, patiently explaining your methods and answering my sometimes-dumb questions.

Kayla Cloonan, my niece, for help with design, illustrations and videos that accompany this book, and for your enthusiastic, continued encouragement, artistic sensibility and so much more.

Barry Reed, my dear friend, for your editing, and perhaps more importantly for your steady, calming voice of moderation that has more or less kept my writing in line.

Julia Rose Golomb, my kind and able daughter, for reading, editing, and providing me with your unique perspective.

Heather Reed McManus for your research and illustrations.

Heather Turner, Joel Levy, Tom Raffio, Dan Feltes, Tom Wessels, Ted Williams and Susanna Berkouwer, Neil Comins and Doug Brooks, for help and counsel in publishing this book.

Thanks to the homeowners who let me follow their home builds, answered my questions, and let me photograph and video them, Mike Marion, Lessa Brill, John Wallace, Andrea and Jeff Burns, Arthur and Debbie Kliman and Jack Bingham.

Thanks to the industry experts whose guidance and knowledge have helped me immensely.

- Martin Orio - Massachusetts Geothermal
- Dana Fisher - Mitsubishi Electric
- Steve Gorse - Home Energy Products
- Eric St. Pierre - Revision Energy
- Aaron Linn - Marvin Windows
- Carl Daniels - Lakes Region Community College

Thanks to the people who supported my GoFundMe efforts to help fund the publication of this book.

Most of all, thank you to my wife, Laurie, and my daughters, Julia and Emily, for years of love and patience with my obsessive motor-mouthing about sustainable energy.

TABLE OF CONTENTS

Introduction 1
Chapter 1
    The Net-Zero Movement Has Begun 5
Chapter 2
    Building Science 11
Chapter 3
    The Envelope, Please 25
RESIDENTIAL RENEWABLE ENERGY SYSTEMS 48
Chapter 4
    Siting For Solar Exposure 49
Chapter 5
    Passive Solar Design 58
Chapter 6
    Solar Hot Water 63
Chapter 7
    Solar Electric (Photovoltaic) Energy 72
Chapter 8
    Air- and Ground-Source Heat Pumps 82
Chapter 9
    The Wallace-Brill Home 94
Chapter 10
    The Kliman Home 108
Chapter 11
    The Burns Straw Bale Home 118

Chapter 12
    The House That Jack Built 131
Chapter 13
    Putting It All Together 143
MARKET TRANSFORMATION: 149
Notes 157
About the Author 160

Dear Reader,

Thank you for purchasing and reading Warm and Cool Homes. It is written for the lay-person so you don't need a background in building or energy to understand the content.

To this end I have also made a series of videos that help to explain the topics. I hope you enjoy the book and videos, and that they provide you with helpful information relevant to your home or future home.

What you see here is a snapshot in time of what I believe to be the best practices for energy efficient building at this writing. However, the field is constantly changing.

For this reason, I felt it was important to have a way to update the book, and you the reader with new information. To access the videos, updates as they come along, and a resource page, please go to www.WarmAndCoolHomes.com.

Thanks again for reading Warm and Cool Homes, I hope you find it interesting and helpful.

Cheers,

Wes Golomb

ATTENTION: FUNDING AVAILABLE FOR ENERGY PROJECTS

Just as we were going to print, the US Senate passed legislation that provides funds and incentives for many of the strategies described in Warm and Cool Homes!

As more details become available, we will post funding opportunities, on the same page as the videos.

Additionally, all of the techniques and technologies that are spoken of in this book can be applied to retrofits as well as new houses. Retrofits will be eligible for funding too.

Please go to WarmAndCoolHomes.com and register. You'll get access to more than a dozen videos, funding opportunities, our resource page and added information about retrofits.

WarmAndCoolHomes.com

INTRODUCTION

I've been obsessed with energy my whole adult life. Early on I understood that increasing the energy efficiency of our homes is the most effective action society can take to save both energy and money. It seemed rather obvious to me. If you have a leaky bucket, rather than turning the water on at a higher flow to keep the bucket full, you need to plug the hole.

Maybe it's because I HATE being cold, or the residual effects of coming of age just as the first oil embargo hit, but I became fascinated with the potential for buildings to save energy and money. In the early 1980s, I did a stint as a salesman for an insulation and window company. They sold their products after doing an "energy audit" on a house. For a brief period of time, New Hampshire certified energy auditors, and one of the requirements for my job was that I get certified. I became the second certified energy auditor in the state. The first was the director of the energy office.

I'm not sure why I needed to get certified. The result of the audits was supposed to be the same for every house: sell 'em windows and insulation. As I began to learn more about what we now call building science, selling windows and insulation—a one-size-fits-all solution—got old fast. I subsequently took a job in a utility-sponsored efficiency program in Massachusetts.

My career path took me away from buildings and energy in 1982 when I got the opportunity to teach environmental science at a local college. In 1999, I found myself back in the energy field again when I took a job administering New Hampshire's Energy Codes for the Public Utilities Commission (PUC). It was my job to inspect and certify that the plans for new structures in the state met the minimum code.

I took every opportunity to talk with builders about energy-efficient building techniques and to be the state's resource for building more efficiently. I was invited to join the board of the New Hampshire Sustainable Energy Association (NHSEA).

My time administering energy codes and serving on the NHSEA board really became the roots of this book and video series. While at the PUC, I met Bob Irving, now fondly known in this book and video series as "Bob the Builder." While most builders thought I was some kind of nut for trying to get them to build better houses, Bob actually changed his business and he now builds only net-zero, energy-efficient homes. In the last three years, he has won first place and third place in New Hampshire's yearly contest for the best-built net-zero home!

Since retiring from teaching, I have followed Bob's net-zero home-building process closely through photos, videos, and interviews.

NHSEA used to run an annual Green Building Open House Tour. On an early fall weekend, homeowners would open their homes to demonstrate a wide variety of sustainable techniques they had used in building, remodeling, or upgrading their homes (and they usually provided hot cider, donuts, and apples).

Through the Green Building Open House, I met Jeff and Andrea Burns, who built the straw-bale home profiled in this book.

This book and video series are, in part, an attempt to continue in the spirit of the Green Building Open House tour. My intent is to lay out the basics of energy-efficient building and retrofitting. Then, rather than having you visit a net-zero or near net-zero home in person, I will do my best to bring them to you in these pages and videos.

I did not realize this when I started writing, but Warm and Cool Homes is also about how people with different lifestyles have built sustainably in ways that match their individual needs. Chapters 1–8 are about the building science principles that are used to build or retrofit homes to be airtight, as well as the renewable energy systems that are commonly incorporated into these homes. Using our current understanding of building science and increasingly cost-effective sustainable energy technologies, it is possible to build energy-efficient houses that can be entirely powered with renewable energy.

Chapters 9–12 show how different people approach the quest to build their dream home with a shared goal: a sustainable house with a limited carbon footprint that meets the needs and lifestyle of its occupants in the context of a changing climate.

In the United States, buildings use almost 50% of the energy we consume. A large portion, as much as 60% of the energy that powers our homes, is needlessly wasted. This massive waste of energy is the lowest-hanging fruit, the most cost-effective opportunity we have for cutting our fossil fuel emissions. The good news is that the same actions that could bring about large cuts in fossil fuel emissions have tremendous co-benefits.

Building energy-efficient homes is labor intensive. As such, building efficiency and renewable energy are among the fastest-growing job sectors, contributing large numbers of jobs and lots of economic activity.

Better health is another co-benefit of energy-efficient housing. A well-built, energy-efficient home is healthier. There are fewer issues with mold, mildew, and air quality. I have interviewed several people who reported abatement and the vanishing of respiratory symptoms after they moved into an energy-efficient home.

Buildings consume over 70% of the electricity we produce, 80% of which is generated by fossil fuels. This partially accounts for the fact that the United States produces 16% of the world's carbon emissions while accounting for only 4% of the world's population. This at a time when climate science tells us we must reduce our carbon emissions immediately! If we want to mitigate the worst effects of climate change, we have a critical role to play in converting to zero-carbon emissions as quickly as possible.

Increasing the energy efficiency of buildings represents a tremendous opportunity. Individuals can save money and live in a more comfortable, healthy environment. But lower utility bills are only the start. Building energy-efficient homes (and other structures) represents an opportunity for local economic growth due to the demand for skilled labor. When these homes are powered by grid-tied solar power, as most are in 2020, they make the whole grid more stable and resilient.

Looking at the big picture, mitigating climate change requires a transition from fossil fuels to sustainable energy. The most effective means for accomplishing this transition appear to be:

> **Energy Efficiency** – Implement energy-efficiency measures that allow us to use only as much energy as is needed to do a particular job.
>
> **Electrification** – Convert engines that run on fossil fuels to motors that run on electricity. This would save approximately one-third of the energy currently consumed.
>
> **Renewable Energy** – Get all needed energy from renewable sources.

This big-picture strategy is also the strategy behind building net-zero homes: make the house as energy-efficient as possible, use only electric-powered energy-consuming devices, and get all needed electric energy from sustainable energy sources, usually solar photovoltaic panels.

Adoption of this strategy could lower the energy consumption of a significant number of homes and would lower greenhouse gas emissions, increase economic activity, and generate tax revenue. These benefits will only increase. As more and more energy-efficient homes are built, like with anything mass produced, our production skills will increase and the costs will decline.

Image 1-1 Fregosi home, Raymond, NH
Designed by Michael Green & built by RH Irving Homebuilders
Photo: Wes Golomb

CHAPTER 1

The Net-Zero Movement Has Begun

No two houses are the same. They range in energy efficiency from extremely inefficient houses to extremely efficient houses that generate enough solar power to provide all the energy the house needs. The latter are called net-zero homes and are the holy grail of residential energy efficiency. There are actually some houses that generate more energy than they consume and have the potential to power an electric car in addition.

As our society begins to understand the environmental costs of fossil fuel energy consumption, there is a growing interest in building homes that use less energy. This book will show you what you need to know to make sure the house you are building or retrofitting is among these efficient, comfortable, and healthy homes.

When we examine the energy consumption levels of new homes, we find a range of efficiency levels, from very inefficient homes that are poorly insulated, poorly air sealed, and sloppily constructed to homes that generate more energy than they consume.

Energy-efficient homes range from slightly-above-average efficiency homes to net-zero homes, the latter of which generate their own energy and do not use outside energy sources. Some homes that use sustainable energy generation, such as solar or wind power, are net-positive—that is, they produce more energy than they consume. This book focuses on building net-zero (NZ) homes and net-zero ready homes. NZ homes generate all the energy they need, and NZ-ready homes are homes that will become NZ with the addition of energy generation means such as solar power.

Over the past twenty years, our understanding of building science has improved, as have the costs and level of available technology. It is now practical and cost-effective to build a net-zero home that generates all the energy it needs for heating, hot water, and electricity. In a recent white paper, the Rocky Mountain Institute concluded that a net zero home's cost differential is about 3% more than a code house. Add solar (electric) to that house and the cost differential jumps to 8%. There are, of course, a range of situations and costs that can amount to a cost differential as high as 15% in my review. However, this is not a huge amount when one considers the eventual savings that the initial investment buys.

According to the Environmental Protection Agency's Energy Star program, the average U.S. household spends more than $2,000 per year on utility bills. Other estimates show that utilities cost upward of $2,500 a year. In the Northeast, when both heating and cooling are included, these numbers rise an additional $1,000 or more.

For the average American, this works out to about 7% of annual income, but for low-income households this can be more than 20% of annual income. The situation is not improving, as electric rates increased by 33% between 2005 and 2016. In June of 2021, in NH where I live, the Public Utilities Commission just approved another 33% increase in rates.

Building a net-zero home can cut these costs by up to 90%. Instead of utility bills costing $2,500 annually, the average utility bill for net-zero homes is in the range of $250 a year.

Three of the net zero homes we will examine in this book pay $400 or less in utilities per year. The fourth home probably would pay that or less except that it also charges the families electric car. Their total bill for heating, cooling, electricity and charging the car was abut $1000 in 2020.

Despite the current positive state of building science, technology, and costs, the average home today barely meets the minimum standards set by state energy codes, which are the legal minimum standard for energy efficiency in buildings. I know this because I spent almost ten years administering the New Hampshire Energy Codes, and I have stayed in close contact with the stakeholders. I regularly talk with builders, building inspectors, and people who are building homes, and I hear their concerns as well as the complaints of the owners who have lived in the houses they built.

The Energy Codes are part of a suite of building codes that aim to ensure buildings are safe and healthy. The Energy Codes define the legal minimum energy efficiency to which a building can be built. Energy codes, particularly in my home state of New Hampshire, are a low bar to meet. At the time of this writing in 2021 New Hampshire's Energy Codes have been updated from the 2009 version to the 2015 version with amendments. These amendments weaken the requirements of the codes and the procedures by which they are enforced.

According to the current national plan, net-zero homes will be in the energy codes within the next few years. This will be good for our future housing stock if states follow through and adopt such codes and then enforce them. Whether this will happen remains to be seen.

The average house uses much more energy than it requires, adding to the estimated 60% of our total energy that is wasted. This is the bad news—we're still building energy-inefficient housing.

The good news, as you will see in this book, is that it is not hard, or particularly expensive, to build efficient, comfortable, and healthy homes that prove to be a smart, cost-effective investment.

Over the next thirty-five years, the United States will add two trillion square feet of buildings. The U.S., which comprises about 4% of the world's population, produces 16% of global carbon emissions. Clearly, we have a duty to play a critical role in responding to climate change; making our housing stock net-zero is one of the most cost-effective ways of lowering our emissions while improving the comfort and health of our homes in the process.

According to Sam Rashkin, chief architect of the Building Technologies Office in the Department of Energy's Office of Energy Efficiency and Renewable Energy, there are currently over two million Energy Star-certified homes and over 2,700 Energy Star builder partners. More than 500 builders built over 5,000 certified net-zero homes in 2019. A recent inventory of housing efficiency in North America done by Team Zero, formerly the Net-Zero Coalition, found over 22,000 zero-energy homes in Canada and the U.S., which represents 400% growth from 2015 to 2018.[1]

Many builders recognize the importance and value of building high-efficiency homes. In a multi-year study, Dodge Analytics asked builders if they have built or plan to build a net-zero home. The percentage of those responding "yes" increased from 21% in 2015 to 44% in 2019, yet another indicator of the growing popularity of net-zero homes.

By 2018, nineteen mayors of large cities across the world (seven from the United States) had signed a net-zero carbon pledge that affirms that all new homes will be net-zero by 2030 and all existing housing will be net-zero by 2050.

During the summer of 2019, one hundred cities across the world committed to net-zero emissions. Nine states have committed to being carbon free by 2050, and six states have already made significant movement toward net-zero energy codes. Prior to 2019, just five countries had committed to net-zero emissions. In 2019, seventy-seven countries committed to net-zero carbon emissions by 2050.

The net-zero movement has definitely begun.[2]

Rating Systems

How can someone tell whether their house meets or exceeds the minimum standard and how close it is to the net-zero mark? Most jurisdictions have an energy code, which is the minimum legal efficiency standard. Don't confuse energy codes with energy-efficient buildings. Energy codes lay out minimum standards, but you're likely reading this book because you're interested in doing better than the minimum.

There are a variety of independent groups that have developed various standards and rating systems to assess the energy characteristics and performance of houses. These rating systems include:

Energy Star – developed by the U.S. Department of Energy

Home Energy Rating System (HERS) – developed by the Residential Energy Services Network (RESNET)

Leadership in Energy and Environmental Design (LEED) – developed by the U.S. Green Building Council[3]

Passive House – developed by the International Passive House Association[4]

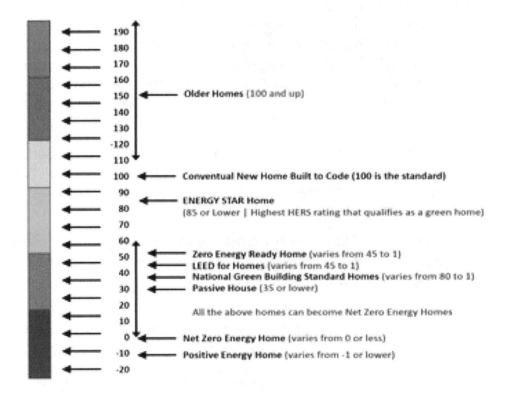

Image 1-2 How common rating systems compare to the HERS scale.
Diagram: Kayla Cloonan

You will also hear talk of net-zero (NZ) energy-ready and net-zero (NZ) homes, which I have already mentioned above. The chart below compares these rating systems.

HERS Ratings

The Home Energy Rating System (HERS) measures a home's energy efficiency. The lower the HERS rating, the more energy-efficient the home. The HERS rating compares the house in question against a reference house with a rating of 100 and a net-zero energy house with a rating of 0. Each number represents a 1% improvement in energy efficiency.

A HERS rating is arrived at through an energy audit and a blower door test, which measures air leakage. Each rating falls on a scale from 200 (worst, a terrible energy waster) to 0 (net-zero energy). The middle of the scale, 100, is the energy code reference home score, that is, the number that would be assigned to a home that just meets the energy code. A home that generates more energy than it uses would get a negative number as a rating. In 2019, more than 40,000 HERS ratings were given, with an average score of 59.

If you were in charge of keeping a leaky bucket full, what would you do? Get a bigger hose, or try to seal the leaks? Sealing the leaks would be more effective. Then, after you seal the leaks, you might consider a more efficient way of keeping the bucket full. So too with buildings. Our first priority is to build (or retrofit) buildings to use less energy. Then we work to generate by sustainable means the reduced energy still needed to power them.

> YOU MUST EAT YOUR EFFICIENCY VEGETABLES BEFORE YOU HAVE YOUR SOLAR COOKIES!

Building a net-zero home requires a different overall approach to building. Traditionally, homebuilding is viewed as a series of unrelated tasks: digging the foundation, pouring the slab, building the frame, and adding insulation, electrical, and plumbing. The result of this sequential process is that houses often have a whole host of issues, including excessive heat loss, air quality problems, mold, ice dams, and/or rotting frame members.

Many of the concepts suggested in this book come from passive house recommended techniques. Sustainable builders approach the house as a system. When a house is viewed in this light, all the tasks listed above are related. For example, there is a connection between how the home is framed and what type and quantity of insulation is used. The type of lighting chosen and how it is installed can greatly affect how much heat loss will occur, as well as the overall cost of heating and lighting. How the house sits on the foundation and the details of how it's put together can help control heat loss as well as mold growth and air quality.

When the goal is an energy-efficient shell, it is key to plan and then build an envelope that is moisture resistant, air sealed, and insulated. The envelope is the surface that separates the inside of the house from the outside, and it includes the walls, floor, and roof.

It is important to plan the location of the vapor (water), air, and thermal (heat) barriers along the same plane and then build them with no gaps in between. The typical weak points in the vapor/air/thermal barriers are found at joints where walls and/or roofs meet. Ensuring that these areas remain insulated and air sealed is key to building a high-performance home. To understand this better, let's take a look at a few key building science principles.

CHAPTER 2

Building Science

HEAT LOSS

Let's return to the leaky bucket analogy. The bucket represents the building, and instead of leaking water, the house is losing heat. When asked, "How does heat move?" many people would answer, "It rises." Actually, cool air is heavier than warm air, so it sinks, forcing the warm air up. However, when we are talking about a solid material like a silver teaspoon or the floor of a house, Heat always moves from warm to cold. If a ground floor room is 70 degrees and the temperature of the earth below is 50 degrees, heat will move from warm to cold—in this case down, through the floor and out of the house, continually.

Heat predominantly moves in one of two ways: conduction or convection. Conduction is how heat moves through a solid material. The classic example is a silver teaspoon in a cup of hot tea. You can easily feel that the handle warms as the heat readily conducts from the tea and through the spoon. Conductivity represents the ability of heat to move through a material. Conductivity is measured in µ-value The higher the µ-value, the better the conductivity.

The efficiency of a window is measured by its µ-value. Because we want as little heat loss as possible, the lower the µ-value the better, as this means less conductive heat loss will occur. Wood does not conduct heat as well as a silver spoon, but it still conducts a relatively large amount of heat. We will soon see why this is an issue.

INSULATION

A material with a high ability to resist the conduction of heat is called an insulator. The measure of an insulator's ability to resist conductive heat loss is its R-value; the higher the R-value, the more resistance to heat a material has.

We can look at the same material from two perspectives: its conductivity (how well it conducts heat) and its R-value (how well it resists conducting heat). Each is a different way of describing the same thing. Mathematically, the R-value is the inverse of the µ-value. Thus, $µ=1/R$ and $R=1/µ$.

Most homes built in the last fifty years were constructed with 2x4 or 2x6 framing. As much as one third of a wall's surface area is made of framing, which conducts heat from warm to cold. This is called a thermal bridge and is responsible for significant heat loss in many homes.

We can visualize heat loss by using an infrared camera that sees heat. Here are two infrared photos of a house, the first (left photo) from the outside and the second (right photo) from the inside.

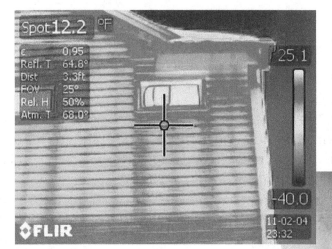

**Image 2-1 Thermal bridging as seen from the outside. The studs are warm (red), relative to the cold nighttime winter air, indicating conductive heat loss.
Photo: Wes Golomb**

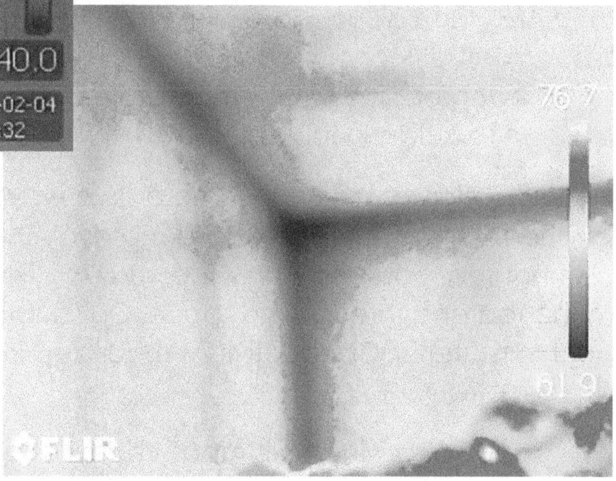

**Image 2-2 Thermal bridging from the inside. The studs, cooled by their exposure to the outside, are cooler than the room temperature and show blue.
Photo: Wes Golomb**

You can see that the range of temperatures represented in the first (left) image is -40 to 25.1°F. This photo was not taken in the Arctic. -40°F represents the temperature of the night sky. Areas of heat loss from the house are represented in white and red. You can clearly see the thermal bridging through the studs where the heat is conducted from the warm inside to the cold outside.

What you see is the result of having insulation, usually fiberglass, only in the bays between the studs. This insulation design leaves a large percentage of the wall's surface area filled with uninsulated wooden studs, which conduct heat around the insulated bays.

To effectively insulate a stud wall, some form of continuous insulation should be used in conjunction with the cavity insulation. Typically, this would be some kind of foam board, but recently rigid rock wool insulation panels have become available.

Over the past fifty years, fiberglass has been the most commonly used form of insulation. In a 2x6 wall, fiberglass can provide at most an R-21 insulation value in the wall cavities. Unless fiberglass is fully lofted and touching all the surfaces of the cavity, it is not an effective insulator. Fiberglass insulates by trapping air and thus slowing down its movement, but it does not stop air from moving through or around the fiberglass.

Consider air and a house as you would water and a leaky boat. It doesn't matter how thick the rest of the hull (or insulation) is; if there is a hole, water (cold air) will leak in, and if it is not stopped, it will sink the ship. Plus, as warm air leaks out, it carries moisture with it.

The following chart describes the characteristics of various sources of insulation. Each insulation type has advantages and disadvantages and (for the most part) an appropriate use.[5]

The Energy Geek's

TheEnergyGeek.org

	R Value Per Inch	Form	Vapor Permeability	Ability to recover From getting wet
Fiberglass	R 3.5/inch	rigid, batts or loose fill	vapor permeable	low
Loose Packed Cellulose	R 3.5/inch	loose fill	vapor permeable	absorbs water but will dry out
Dense Packed Cellulose	R 3.7/inch	loose fill	vapor permeable	absorbs water but will dry out
Rock Wool	R 4/inch	rigid, batts or loose fill	vapor permeable	water drains through it
EPS Foam Board	R 4/inch	board	vapor permeable without foil, Impermeable with foil	doesn't absorb water
XPS Foam Board	R 5/inch	board	moderate	doesn't absorb water
Polyisosanurate Foam Board	R 6.5/inch	board	impermeable	absorbs water
Single Part Open Cell Spray Foam	R 3.5-3.6/inch	spray foam	vapor permeable	doesn't absorb water
Two Part Closed Cell Spray Foam	R 6.5/inch	spray foam	vapor permeable	doesn't absorb water

Image 2-3 Insulation Fact Sheet
Chart: Wes Golomb and Kayla Cloonan

Insulation Fact Sheet

Air Barrier	Global Warming Factor	Ease of Installation	Appropriate Use	Relative Cost	Other
no	high	easily installed but rarely installed properly	sound deadening between interior walls, secondary insulation with foam	low	should not be installed in exterior walls
fair	low	blown in	attics	higher than fiberglass	treated with Borax as a fire retardant
yes	low	blown in under pressure	walls and other contained cavities	higher than loose pack cellulose	treated with Borax as a fire retardant
yes	moderate	easily installed	any envelope surface	higher than fiberglass but relatively low	Non-flammable
yes	lowest of rigid insulation board	easily installed	thermal break, good in damp places	cheapest of foam boards	do not use with foil on the outside of a house
yes	high	easily installed	thermal break	higher than EPS	different color per brand - Green, Blue, Pink. Also known as Styrofoam
yes	moderate	easily installed	thermal break	similar in price to XPS	minimum 2" on exterior walls to avoid moisture Condensation
yes	high	spray can (can be installed by consumer)	filling small holes	high	used with gun for small areas
yes	high	large amounts come on a truck and must be professionally installed	can be sprayed on any surface needing insulation	high	may be issues with health effects of outgassing chemicals. Expensive

Here are four examples of ineffective insulation installations:

Image 2-4 Gaps allow air and heat to flow around insulation.

Compression makes it ineffective.

Most houses are insulated in the bays between studs with fiber insulation (cellulose, fiberglass and rock wool). This strategy leaves the studs as thermal bridges which conduct heat from the house in winter and into it in the summer.

Image 2-5 Pipe compressing insulation makes it ineffective

Images 2-6 and 2-7 Holes in insulation coverage, like holes in the hull of a boat, make these insulation jobs useless.
Photos: Wes Golomb

A sheet of continuous rigid insulation, on the inside or out can act as a thermal break and greatly limit the conductive heat loss.

If a wall gets wet it needs a way to dry out, so take care not to inadvertently add rigid insulation, which will act as an air barrier, to a wall system which already has an air barrier on the other side.

CONVECTION AND INFILTRATION

Convection is the other predominant way in which heat moves. Convection is the source of wind. For example, when the land is warmer than the ocean, an offshore wind of cooler air moves inland from over the ocean to replace the warm air, which is forced up. During hours of sunlight, water absorbs more heat than land does. After sundown, the water becomes warmer than the air, forming a convection loop in the other direction that causes the wind to blow off the land.

Image 2-8 Offshore breeze
Image: Kayla Cloonan

The same phenomenon is true in your house and even in your walls if they are not well insulated. It goes like this: cold air leaks into, or infiltrates, the living space of a house through any holes. As a result, the temperature drops and the heating system comes on to warm the cold air. The newly warmed air is displaced by more cold air coming in, so it rises. The warm air then either moves out through holes in the envelope or cools and sinks. This pattern causes a never-ending convection loop—fueled with your dollars—by constantly heating cold air that is replacing the warm air that has leaked out!

Image 2-9 Onshore breeze
Image: Kayla Cloonan

Drivers of Convective Heat Loss

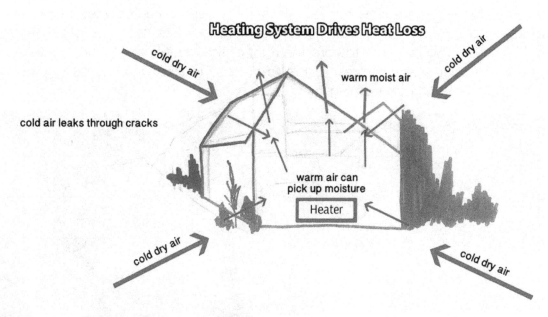

Image 2-10 Heating system drives convective heat loss
Image: Kayla Cloonan

On a large scale, we call this moving air wind. In your house, you call it drafty.

Air infiltration via convective loops driven by heating systems is often the largest source of heat loss and wasted energy in a building.

Image 2-11 A convective loop in an uninsulated or poorly insulated wall
Image: Kayla Cloonan

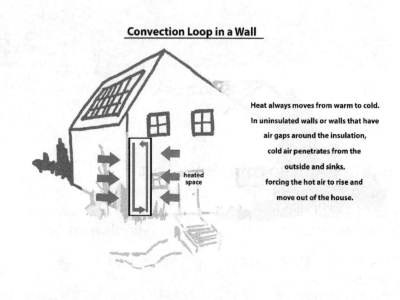

The same process happens on a micro scale in portions of outside wall cavities that are left uninsulated. Holes allow cold air to leak in and sink, forcing warm air up within the wall cavity and thereby forming a convection loop (again fueled by your heating system). That loop pushes heat out of the building and drives your heating bill through the roof.

In Image 2-12, which captures heat loss through a ceiling, you can see both conductive heat loss and convective/infiltrative heat loss. Conductive heat loss (dark blue) occurs through the framing member toward the right of the image, while convective/infiltrative heat loss (the blue wispy shapes) is due to cold air leaking into this ceiling from cracks along the unsealed wood framing.

Image 2-12 Infiltration into a ceiling

Photo: Wes Golomb

Stop the Air, Stop the Heat Loss

To stop the convection loop, an effective air barrier is required on all surfaces that separate conditioned (heated or cooled) spaces from the outside. In standard houses, air barriers are often non-existent or ineffective, as demonstrated by the following images.

Lack of air sealing along framing where different materials meet causes extensive infiltration:

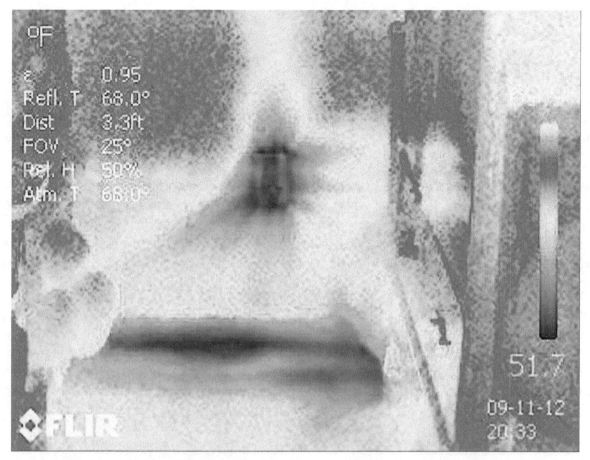
Image 2-13 Conductive and infiltrative heat loss around an electric socket and floor-wall connection
Photo: Wes Golomb

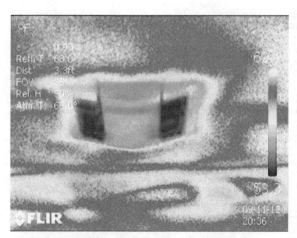
Image 2-14 Cold air leaking in through a bathroom vent
Photo: Wes Golomb

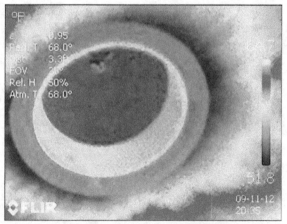
Image 2-15 Infiltration around a recessed light
Photo: Wes Golomb

HEAT LOSS AND MOISTURE

The Mayo Clinic says we need between 30% and 50% humidity to be comfortable in an indoor environment. Too little moisture in a home, or moisture in the wrong places, can make our homes uncomfortable and unhealthy. Air sealing a home helps to keep the humidity at a comfortable level. Here's how.

Have you heard the term relative humidity (RH)? Relative does not mean that moisture is a second cousin to wind. The "relative" in RH refers to the fact that the amount of moisture that air can hold is relative to the temperature. Warm air can hold more moisture than cold air. If we took a box of air at 30 degrees and 99% humidity, and another box of air at 90 degrees and 99% humidity, and wrung them both out, we would find that the warm air contained a lot more water.

As we have seen, the heating system in a leaky house causes a convective loop that constantly sucks cold air into the home. Since cold air cannot hold a lot of moisture, it tends to be dry. When cold dry air moves into a house and is heated, it becomes warm dry air just waiting to absorb moisture in the house. If there is a convective loop, the now-warm moist air rises and escapes through the cracks in the walls and ceiling, carrying off the moisture it captured.

Image 2-16 Heating system diagram
Image: Kayla Cloonan

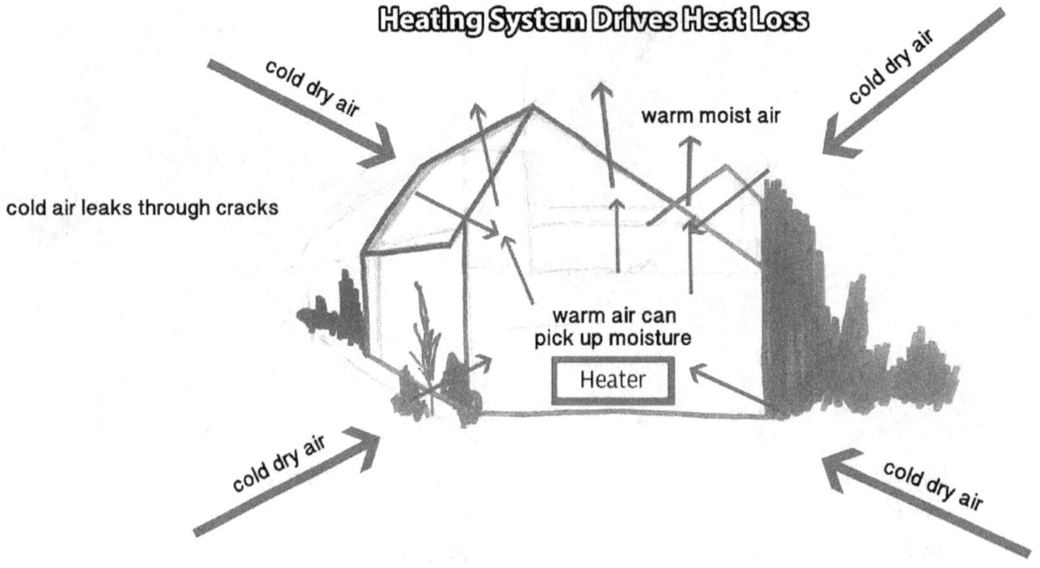

My house, built in the 1980s, often had a relative humidity of 10% or less in the winter until I did a retrofit. After air sealing the house, the average humidity remains above 30%. I now use less fuel to heat my house, and it is far more comfortable in the winter than it previously was.

When a house is not well insulated and well air sealed, condensation can become a moisture problem. When moist air cools, it condenses, thus depositing water on the cool surface. This is what causes dew on the grass on an early summer morning. If warm moist air comes in contact with a cold surface in your house, like a cold roof or any other surface that is below the dew point temperature, it will condense. This moisture can cause wood to rot and mold to grow. Because fiberglass, like a sweater, allows air to move through it, the dew point—the temperature at which moisture condenses on a cool surface—can be reached inside a wall cavity. When moisture condenses on insulation, it can cause that insulation to fail. When moisture condenses on wood and other natural surfaces, it provides the right environment for mold growth: mold spores everywhere, an organic substance (wood) for mold to live on, and the water that condenses on the wood.

Ice dams are another issue brought on by infiltration convective loops. When warm indoor air hits a cold roof, it melts the snow, which runs down as water until it reaches a cold portion of the roof such as the soffit, where it refreezes. This causes ice dams, which are a major source of roof damage and failure because they prevent melting water from shedding off the roof. Instead, melting water backs up into the house.

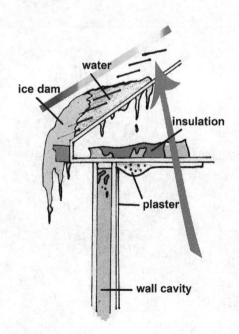

Image 2-17 Ice dams form wh[en] escapes the envelope and me[lts] the roof.
Image: Kayla Cloonan

CHAPTER 3

The Envelope, Please

The Marion House

About twenty years ago, while I was administering the New Hampshire's Energy Codes, Bob Irving, a local home builder in central New Hampshire, brought me a plan for energy code approval. It was another of the many houses I saw that, on paper at least, met our minimum energy code. I gave him my standard spiel: He could make some inexpensive, incremental improvements that would in turn make the house more energy efficient. Unlike most of the people with whom I spoke, Bob not only listened to my recommendations, but then made building energy-efficient homes the center of his business. Now he exclusively builds net-zero and net-zero ready homes. In 2018 and 2019, Bob won statewide awards for the net-zero homes he built in New Hampshire. I've watched Bob and his crew build four net-zero homes.

Though each home is different, all of Bob's houses use the same fundamental building science concepts we've just discussed. This next section describes how those concepts were applied in building a home for Mike Marion in Newmarket, New Hampshire. Mike's net-zero ready house was completed in late 2017 and includes thirty-six photovoltaic panels that were installed in May 2018.

The house is constructed with double walls to avoid conductive thermal bridging and is meticulously air sealed to avoid infiltration.

Opposite page:
Image 3-1 The south (business) side of the Marion home.
Photo: Wes Golomb

Image 3-2 Drains are placed in the gravel, and the foundation is waterproofed and then backfilled with gravel.
Photo: Wes Golomb

Moisture control begins with assessments of the soil and drainage patterns. After the foundation is poured, drains are dug and placed in gravel. Before the foundation is backfilled, the drains are covered with a layer of landscaping cloth to keep dirt from clogging them. This ensures that the basement will stay dry and that water will move away from the foundation.

The foundation of this house was poured using aluminum forms, which support the drying concrete and leave the surfaces smoother than those made with wooden framing. The smooth surface allows the rigid foam insulation to adhere flat against the concrete, thereby limiting air gaps that reduce its ability to insulate.

Image 3-3 Foundation ready to be backfilled
Photo: Wes Golomb

Image 3-4 Basement windows are double glazed.
Photo: Wes Golomb

Standard houses typically use single-pane windows in the basement and double-pane windows in the interior of the structure. In Image 3-5 you can see the double-paned windows installed in the foundation.

Manufactured trusses are built with short pieces of wood and reinforced. There are two advantages to trusses: (1) Without the need for structural walls, the floor size can be larger, thus allowing for a much more flexible floor plan, and (2) there is no need to cut down the largest, oldest trees.

Image 3-5 Manufactured trusses
Photo: Wes Golomb

Image 3-6 Double wall between living space and garage
Photo: Wes Golomb

Image 3-7 Outside double wall with framing for window
Photo: Wes Golomb

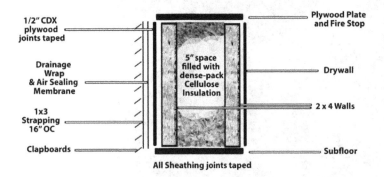

Image 3-8 Double wall assembly
Image: Wes Golomb

Remember the infrared photos of the walls conducting heat to the outside? In traditional houses, this can be a significant source of heat loss. One way to minimize conductive heat loss is to build two separate walls that are not connected. The space between the stud walls is called a thermal break because it limits thermal bridging to the plywood plates at the top of a 12-inch cavity, which will be densely packed with cellulose insulation.

Image 3-9 Douglas fir wall sheathing
Photo: Wes Golomb

The outside of the structure is designed to keep water away from the wall and allow any water that may get in to dry to the outside, that is, to drain into the open air and not into the building. Overhangs move water away from the walls. The plywood in Image 3-9 is Douglas fir, which is more moisture-resistant than oriented-strand board.

The roof is made from a product called Advan Tech, which is also often used for subfloors because of its moisture-resistant capabilities. In Image 3-9, you can see that the HydroGap installation has begun on the triangular wall. HydroGap has a rough, bumpy surface that allows for space between it and the plywood in order to further divert water away from the house. The rest of the walls will be covered with HydroGap. HydroGap is used to ensure that the sheathing stays dry. It is not an air barrier.

Image 3-10 An EDPM rubber gasket seals the sill.
Photo: Wes Golomb

AIR SEALING

Most homes lose a significant amount of heat through gaps in the sill, where the walls meet the concrete. Other places are also prone to infiltration. Examples include walls and windows, where utility services enter and exit, around unsealed recessed lights, and through other holes in the envelope. To ensure a tight house, every joint and every surface where two pieces of material come together is air sealed.

Because concrete and wood expand and contract at different rates, the point where they meet at the sill (where the wooden structure sits on the concrete) is often a major source of air infiltration. To avoid this, a rubber gasket is placed at the sill. This gasket extends around the entire perimeter of the house.

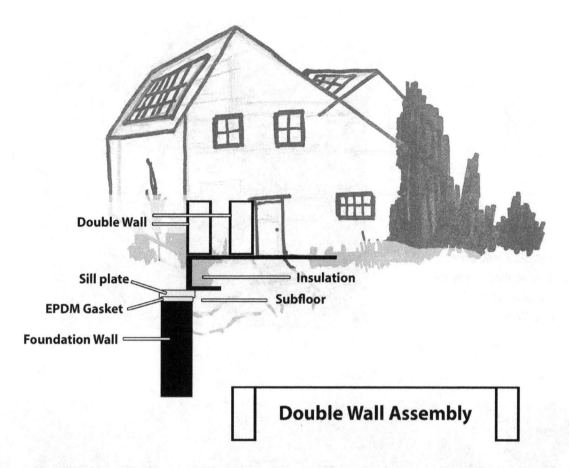

Image 3-11 Floor assembly diagram
Image: Kayla Cloonan

The envelope of this house also includes the basement. The concrete floor is poured onto four inches of rigid polystyrene (coffee cup) insulation. Similarly, two 2-inch layers of rigid foam polyisosanurate insulation cover the basement walls. The top layer is rated for indoor living and is painted with fire-retardant paint. Above the walls with rigid insulation, the sill is insulated with spray foam that has a high R-value and also provides an effective air barrier around the sill.

The combined wall and sill insulation gives an unbroken layer of R-24 insulation, separating earth from the conditioned space and also effectively air sealing the whole basement and sill area. For context, a typical 2x6 stud wall with well-installed fiberglass conducts a lot of heat through the uninsulated studs, bringing the overall R-value of the wall to R-15 at best.

WARM AND COOL HOMES

Image 3-12 Insulation and air sealing of the sill and basement walls
Photo: Wes Golomb

Opposite page:
Image 3-13 All joints between sheathing are taped to limit infiltration
Photo: Wes Golomb

Image 3-14 Details of air sealing at joints and around windows and at sheathing joints
Photo: Wes Golomb

All sheathing and joints at doors and windows are taped to air seal any leaks. This forms the air barrier. HydroGap, a drainable housewrap, is then applied to assist in moving water away from the house. HydroGap is designed to leave a 7 mm gap between it and the siding. This gap helps moisture drain away rather than becoming trapped and causing problems.

Since this writing, Bob has begun experimenting with a new peel-and-stick air barrier called Henry Blueskin that takes the place of both taping and HydroGap. The cost will be higher than tape and HydroGap, but it is expected to save a lot of time, because taping every joint is a very time-consuming task.[6]

Image 3-15 Double wall filled with 12" of cellulose insulation (R-42) prior to sheet rock
Photo: Wes Golomb

Twelve inches of cellulose insulation is blown into the double-wall cavities. In Image 3-15, you can see electrical wires coming through the wall. Care is taken to seal around the wall and ceiling penetrations so there will be no air leakage through cracks between the electrical box and the envelope.

How well a home is air sealed is not something that can generally be seen with an unaided eye, but it can be measured with a blower door. A blower door is a powerful fan that fits into an exterior doorway. It is attached to a manometer, which measures the difference in air pressure between the inside and outside of the house.

When the blower door is turned on, air is blown out of the house and the house is depressurized to a standard difference of 50 pascals between the inside and outside. A pascal is a unit of air pressure. By measuring the volume of air that flows through the fan at this standard pressure difference of 50 pascals, we can measure how tight the house is.

Image 3-16 Cellulose insulation is blown into both flat (R-66) and sloped (R-64) roof.
Photo: Wes Golomb

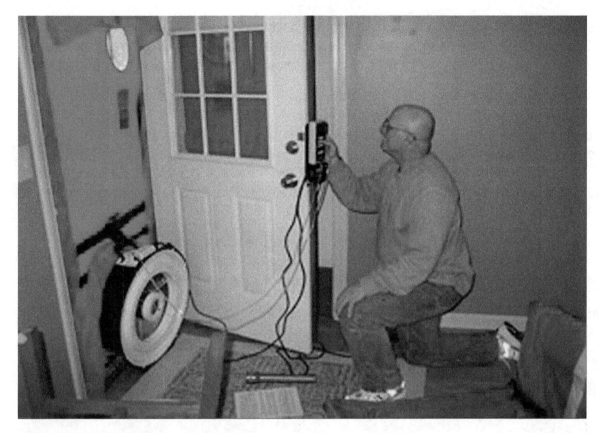

Image 3-17 Blower door and manometer
Photo: Wes Golomb

Try this experiment in your mind. Imagine sucking on a straw and remember how easy it is to draw air through. It would be almost impossible to build any pressure in the straw. Now imagine you're drinking a frozen drink, which mostly blocks the straw. Imagine trying to draw air through the straw. What would happen? It would take a lot more effort to draw air through, and you could easily build pressure in the straw.

The unclogged straw represents a house that is not air sealed. The clogged straw represents an air-sealed house. If the house is very tight (without a lot of leaks), we will be able to depressurize it easily with a small amount of air getting sucked through. If there are lots of holes in the house, like with the unclogged straw, the fan will pull a lot more air through to achieve the standard pressure.

If the fan cannot pull a lot of air through, the house is well air sealed. If lots of air is pulled through, we know that the house is not well air sealed. There is a mathematical formula, which I won't bore you with, that converts the amount of air pulled through a house to a unit called air changes per hour (Ach/Hr). This is the final measurement we get from a blower door test.

In New Hampshire, 7 Ach/Hr is the minimum standard set by the energy code for new construction. In Massachusetts, the standard is 3 Ach/Hr. Bob Irving told me that he aims to keep his numbers below 1 Ach/Hr in the net-zero and net-zero ready homes he builds.

In Image 3-17, you can see both the blower door and a manometer mounted on the door. A manometer measures pressure differences between the inside and outside of the house. The blower door is turned on and sucks air out of the house. It also measures the volume of air that is pulled through the fan. The volume of air indicates the tightness of the house. The unit of measure for tightness is Ach/50, which stands for air changes per hour with the blower door on, and a difference in pressure of 50 pascals between the inside and outside.

In this state, with the house closed tightly and the blower door on, the house is depressurized to 50 pascals less pressure than the outside air. The Marion house turned out to be .9 Ach/Hr, which is an excellent number for a net-zero home!

Here are some comparison points to help interpret Ach/Hr readings: A 2002 study of twenty-four new Wisconsin homes showed a median air leakage of 3.9 Ach/Hr. New home builders in Minnesota routinely achieve 2.5 Ach/Hr.[7] The Canadian R-2000 program has an air-tightness standard of 1.5 Ach/Hr. The Passive House air-tightness standard—a tough standard to achieve—is 0.6 Ach/Hr.[8]

VENTILATION

One of the most common, most logical questions asked after a discussion of air tightness is, "Doesn't a house have to breathe?" The answer is NO! Human beings and other living things within the house need to breathe. This means that the house needs adequate ventilation. With COVID-19 in the picture, proper ventilation is more important than ever.

A house with 7 Ach/Hr certainly has adequate ventilation and more—way more! As we've seen, that "more" steals the heat and humidity from the house. What is actually needed is controlled mechanical ventilation, set to a standard that meets the needs of the occupants of the home.

Current American Society of Heating, Refrigerating and Air-Conditioning Engineers (ASHRAE) standards call for one-third of a home's air volume to be exchanged every hour. In an air-tight home, some form of mechanical ventilation is needed to accomplish this.

A heat recovery ventilation (HRV) system removes stale air from the living space. Before the air is exhausted to the outside, it flows through a heat exchanger that removes some of the heat from the outgoing air. This recycled heat is then used to preheat the incoming fresh air. By maintaining this complete volume of air replacement every three hours, the system ensures occupants get a healthy air exchange without the heat loss that would otherwise be associated with it.

tem Heat recovery ventilation system
Photo: Wes Golomb

Image 3-19 The inside of a heat recovery ventilation system
Photo: Wes Golomb

WINDOWS

Windows are often a major source of heat loss in traditional housing. Some have gone so far as to say that a window is a hole in the wall that you put up with so you can get some natural light into the house.

Choosing the right windows and installing them correctly makes a big difference in the comfort and performance of a house. We've already discussed conductivity, measured in µ-value, which represents how much heat can flow through a material. The New Hampshire Energy Code calls for a window to have a µ-value of 0.32 or below. You'll recall that the R-Value used to measure insulation is the inverse of the µ-value. So a 0.32 µ-value corresponds to a 3.125 R-value. For comparison the same code requires an R-19 or 21 in the walls surrounding those windows. .

There is another way that heat is lost from a building that we have not yet discussed: radiation. Radiant heat flows from a material that has more heat to one that has less heat. You know of radiant heat because it heats directly. A woodstove is a good example—you feel the heat radiating directly from the stove. A good hint that heat is radiant is when it heats your front but not your back.

People, furniture, and walls all store heat. The inanimate objects absorb heat from the environment and humans generate heat and regulate their temperature (98.6).

As a result a warm object (or person) will radiate heat toward the cold glass. Remember heat always moves to cold. This is why you feel cold when you sit next to a mediocre window in the winter even if the window is well air sealed. Radiant heat loss through a window can make a room feel uncomfortable.

While at the PUC, for almost ten years, I kept records of every house I approved while doing energy codes. I found that the average percentage of window surface area in a house built in New Hampshire over that time period was about 13–15% of the wall area.

With this amount of exposure to the outside, the quality of window frame and glass makes a huge difference. What makes a quality window? There are a number of factors.

The first factor that affects the quality of windows is how many panes they consist of. A single pane of glass has a µ-value of about 0.1. A second pane of glass in a window adds another 0.1, but there is also an insulating value to the air trapped between the two panes. The double- or thermal-pane window is the most commonly sold window today. A third pane makes a window even more resistant to heat loss. A triple-pane window is more expensive, but its incremental cost over a double-pane has come down significantly, it is far more comfortable, and it usually pays for itself over a short period of time.

I measured the inside surface temperature of single, double, and triple-pane windows with an outside air temperature of 25 degrees. The surface temperature of the single-pane basement window with a metal frame was only 5 degrees warmer than the outside, which was 30 degrees; the double-pane window was 54 degrees; and the triple-pane window was 70 degrees (room temperature!). It is no wonder the owner described the house as "always comfortable."

A second factor in the quality of windows is the gasses that can be placed between the panes of windows as a replacement for air. Inert gasses like argon and krypton (don't use if you're friends with Superman) are pumped into these spaces. They serve to cut down on mini-conduction loops that would otherwise form within the air gaps. These gasses make windows less vulnerable to conductive heat loss.

A

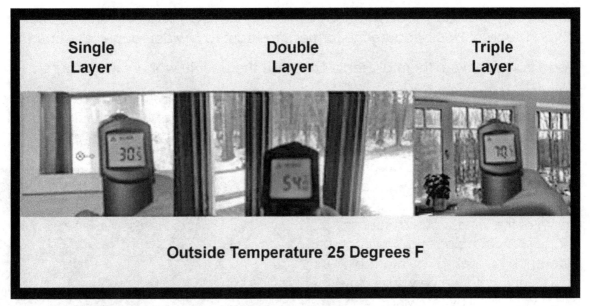

Image 3-20. | Comparison of single-, double-, and triple-pane windows
Single μ=0.55 Double μ=0.35 Triple μ=0.21
Image: Wes Golomb

pane of glass allows a certain amount of light energy through and, when untreated, allows a similar amount of heat energy out. The difference in energy types is because sunlight is shortwave radiation. When a material like a wall or floor absorbs and re-radiates that energy, it does so as heat, which is a longer-wave form of radiation.

A low-emissivity (Low-E) coating may be applied to glass in order to allow shortwave sunlight through while limiting long-wave heat radiation's movement out through the glass, thus adding to the window's efficiency.

A fourth important factor for window quality is the material used for frames. The trick is to find a material that will give the frame strength but not conduct a lot of heat.

Case in point: the metal basement window. The metal frame is highly conductive and is in large part the reason the surface temperature of the window is below freezing. Metal is structurally sound, but it conducts a lot of heat and doesn't make an energy-efficient frame for a window.

Wood does not conduct as much heat as metal, but as we have seen in our thermal bridging discussion, it certainly conducts enough to cause significant heat loss.

Composites are made from wood products and polymer plastics. They have similar structural and thermal properties to wood but last longer.

Fiberglass is easy to make and stable. The frame can be filled with insulation, giving it good thermal properties.

Vinyl frames are made from extruded plastic and they vary in quality. The best frames use uPVC instead of PVC because the former is resistant to a wider range of temperatures.

Good frames have a thermal break, not unlike the double walls in Bob's houses.

Finally, and perhaps most importantly, windows need to be installed correctly. Key to this is integrating the window into the envelope. This involves keeping water flowing out and away from the window, and air sealing around the window with caulking.

The biggest visual difference between standard houses and houses with double walls is the depth of the windowsills, as seen in Image 3-21 The biggest non-visual difference is feeling comfortable on a very cold or very hot day when you stand by the window.

Image 3-21 Net-zero homes use extremely efficient triple-pane windows.
Photo: Wes Golomb

ELECTRICAL SYSTEM

You'll recall that I suggested a systems approach in Chapter 1 when I talked about building a net-zero home. The electrical system is part of that approach. In traditional buildings, it is not unusual for an electrician to come along after the house has been insulated and move insulation, drill through a wall, or otherwise make little changes that not only let electric wires pass, but also let our old friends—air and moisture—get through. Electrical work needs to be scheduled and completed as an integral part of the construction process, without sacrificing air changes. This means working with the electrician throughout the process and identifying places where air sealing is needed after the electrical work is completed.

LIGHTING

All lighting in the Marion house is LED, which saves as much as 90% of the energy consumption as compared to traditional incandescent fixtures. Recessed lights in traditionally built houses generate heat, which, when combined with a lack of air sealing, causes them to be a convective engine, as we saw in Image 2-15. All penetrations through the envelope in the Marion house are well sealed. Because LED lighting gives off a fraction of the heat of incandescent lighting, the convective engine seen in traditional installations is avoided.

SOLAR ENERGY

The house's electricity is generated by 10 kW of photovoltaic (PV) panels on the roof. The size of the PV system was determined by doing a heat loss calculation, which gives a good estimate of how much energy will be needed for heating. In addition, an electric load calculation was done, and the combination of the two calculations gave a good idea of the likely demand for electricity. This is one of the "fuzzier" calculations because of the wide range of demand for electricity due to differences in lifestyle. Based on the first year's data, despite the fuzzy calculation, the PV system was properly sized. In the next chapter, we will discuss solar energy in more detail.

HEATING SYSTEM

Image 3-22 Mitsubishi mini-split inside wall mount
Photo: Wes Golomb

The heating system is a Mitsubishi mini-split. This is a heat pump, an extremely efficient method of both heating and cooling a house. Heat pumps work similarly to refrigerators and will be discussed in detail in Chapter 8. Heat pumps take one unit of electrical energy and convert it into three or four units of heat energy.

WARM AND COOL HOMES

This is particularly effective when solar electricity is used to supply the needed electricity. In this case, one unit of clean solar electric energy can be leveraged to three or four units of heat energy. In Chapter 12 we will look at a house that uses heat pumps in a greenhouse. When the temperature of the greenhouse is in the 80s (think a sunny February day) the heat pumps can leverage as much as three times the input solar energy.

Image 3-23 Mitsubishi mini-split outside unit
Photo: Wes Golomb

Image 3-25 A two-stall garage
Photo: Wes Golomb

Opposite page:
Image 3-24 Front with farmers porch
Photo: Wes Golomb

As you can see from the photos, at my last visit, Bob's crew was finishing up work on the exterior. The interior was complete and was already a comfortable home for Mike and his family.

Image 3-26 Walkout basement
Photo: Wes Golomb

USING THESE TECHNIQUES WORKS

Just before publication I checked back with Mike. He had collated his bills and reported the following:

> We moved into the house in October of 2017 and installed the PV system in May of 2018 so we have a pretty good baseline for that first winter (2017/2018).
>
> First year (winter 2017/2018) - $1542.16
>
> A decent baseline and testament to the fundamental quality of the construction of the house. I think most people would be elated with a total annual utility bill in this range.
>
> Second year (winter of 2018/2019) - $448.63
>
> With the system installed and online in May we missed some energy production but still a good number.
>
> Third year - (winter of 2019 - 2020) - $376.93
>
> Fourth year - (winter of 2020 - 2021) - $347.93

Image 3-27 Dining area
Photo: Wes Golomb

Even in late fall, the windows offer a comfortable, airy space.

The open concept space offers a warm and homey atmosphere. All appliances are energy efficient.

Image 3-28 Bedroom
Photo: Wes Golomb

Image 3-29 Main room
Photo: Wes Golomb

Net Zero Home
MARION

Heated Space: 2600 Sqft
Heat Source(s): Air to Air Heat Pump

Walls
Construction: Double Wall
Insulation: Cellulose
R Value: 42

Foundation: Slab
6" explanded polystyrene
R Value: 24
Basement Walls:
4" polyisosanurate foam
R Value: 28

Roof
Construction: Manufactured Truss
Insulation: Cellulose
R Value: 62

Ventilation: Mechanical

Window Type: Triple Pane
Insulation: Foam/Gasket
U Value: 0.19

Ratings
Blower Door Ach/Hr: 0.9
HERS Rating: 3
PV: 10 kW

Annual Utility Cost
$348

Built 2017

RESIDENTIAL RENEWABLE ENERGY SYSTEMS

The next few chapters focus on the renewable energy systems used in the various other houses profiled in this book. The goal is to provide enough information to help you decide which of these technologies you might want to incorporate into your home. These include solar electricity (photovoltaics), solar thermal (hot water), passive solar design, heat pumps, and geothermal systems.

CHAPTER 4

Siting For Solar Exposure

All the energy humans use can be traced back to the sun, the ultimate source of our energy. Whether we realize it or not, we all experience the sun's force on a daily basis. The melting of ice cubes in our drink, getting into a hot car, and the ache of a painful red sunburn each demonstrate this power that is invisible to the naked eye, unlike fossil fuels, which we can touch and smell. However, it is possible to transform the sun's energy into our primary direct energy supply.

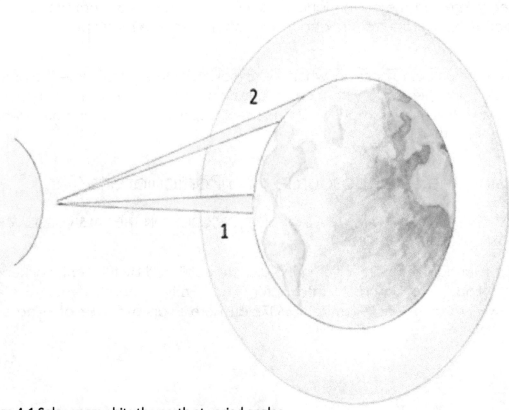

Image 4-1 Solar energy hits the earth at varied angles.
Illustration: Heather Reed

The most direct radiation occurs when the sun is overhead, because the rays are less scattered and less reflected by the atmosphere. As you might expect, the least direct radiation occurs when the sun is on the horizon.

When the sun is directly overhead, the radiation travels a relatively short distance through the earth's atmosphere to reach the surface (1). When the sun is low in the sky, as represented by the angled position (2), the radiation has further to travel through the atmosphere and therefore scatters more energy by the time it reaches the earth's surface.

Through ingenious human effort, we have learned several ways to use this solar energy to our advantage. They all have one rather obvious thing in common: the more sunlight that falls on the solar energy system, the more usable energy the system will generate. Not all potential solar sites are created equal, so before we examine the specific technologies designed to collect and use the sun's energy, let's take a look at how to determine if you have a good potential site.

The sun, which is expected to shine for another five billion years, is the virtually unlimited, sustainable source of energy for all life on earth. About 70% of the sun's radiation that strikes the upper atmosphere is absorbed by the earth. As a result, every fifty-five days, the earth's water absorbs and stores as much heat energy as the planet's known oil and gas energy reserves! The earth's land surfaces also absorb heat from the sun. Over the entire earth, this averages out to about 164 watts per square meter per twenty-four-hour day.

This is a great number to consider when we're looking at the big picture—the potential of solar—but it has little to do with the real question that a homeowner or home builder needs to ask: How much sunlight will hit the specific spot on the roof or ground where you want to collect sunlight?

Assessing the Solar Resource at a Particular Site[9]

THIS SECTION BASED ON ED MAZRIA'S METHOD DESCRIBED IN THE PASSIVE SOLAR ENERGY BOOK. USED WITH PERMISSION.

Detailed solar data for many cities in the U.S. are published by the National Renewable Energy Lab. To find data for a city near you, refer to http://www.nrel.gov/docs/legosti/old/5607.pdf. Here is an example of a data set for Concord, NH.

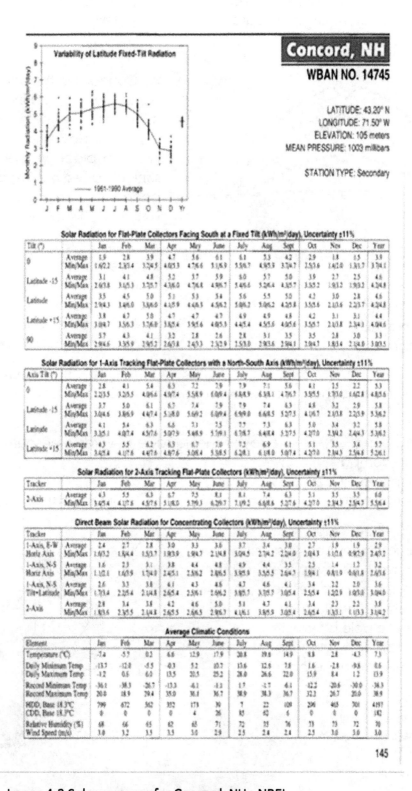

Image 4-2 Solar resource for Concord, NH - NREL

This data set, measured monthly in Concord, NH (the closest location to my house), shows the number of kilowatt hours per square meter per day (kWh/m2/day).

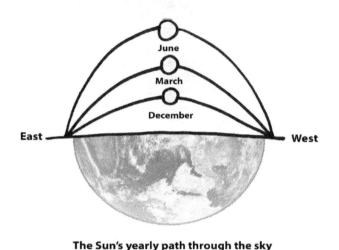

Image 4-3 Seasonal path of the sun
Image: Wes Golomb

As Concord is thirteen miles from my house, it is a good indicator of what I might find with a good southern exposure. However, just because you live in a place where there is lots of sunlight does not mean you have a good site for solar. My house is a great example. I'll show you what I mean.

As I look out my window over the span of a year, I can observe that the sun takes a different path across the sky as the seasons change. The sun is higher in the summer, and as I write this in December, its arc is lower.

To assess a site, we must first determine where solar south is and then analyze what obstacles exist between sun's path and the site.

To get the most out of any solar array, it should face toward solar south. This is NOT the south on your compass. In the northern hemisphere, a compass points to magnetic north.

Over geologic time, magnetic north—the direction a compass points—has moved all over the earth's surface. It is a coincidence of time and physics that at this time the magnetic pole is close to the point we call the North Pole, the point around which the earth rotates.

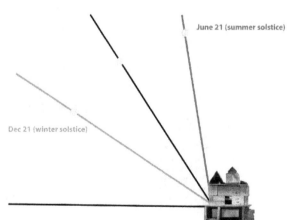

Image 4-4 Seasonal altitudes of the sun
Image: Kayla Cloonan

As a brief aside, geologists have been able to track the path of the magnetic north pole over time by observing the direction of magnetite, a mineral found in lava. When lava is molten, the crystals within magnetite line up, pointing toward magnetic north. When the lava hardens, we have a permanent record of where the magnetic north pole was at the time the magma was deposited. Find out the age of the rock, and you know where the magnetic north pole was at that time. Now back to your regularly scheduled chapter!

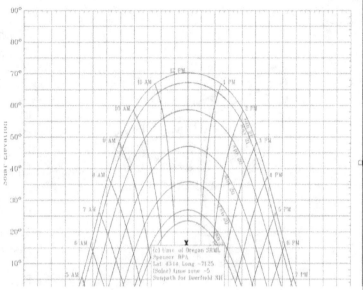

Image 4-5 Sun Chart for plotting solar path and obstacles
Used with permission of Oregon Solar Radiation Monitoring Laboratory.
http://solardat.uoregon.edu/SunChartProgram.html

The angle between the (true) North Pole and the magnetic north pole is called declination. At my house in New Hampshire, that angle is 14°.

True south (180°) is therefore 14° degrees off solar south. 180° + 14° = 194°.

A simple Google search will tell you what the declination at your location is.

After you've determined where solar (true) south is, the next question is what obstacles lie between you and the sun (trees, buildings, etc.) and when and how much they will limit the solar resource. There are computer programs that will calculate this automatically. However, if you don't want to buy a program, you can do it the old-fashioned way by making a chart showing the sun's path and plotting onto that chart the obstacles that will shade a potential site. To do this accurately, you must use two-dimensional measurements. For horizontal (left–right) measurements, we use solar south and total degrees east and west of solar south. At my house, that is 194 degrees.

For the vertical measurements, we use degrees above the horizon. The blue lines on this chart indicate the path the sun takes at my house over the course of the year. Note that the sun path for December through June is marked. As the sun gets lower in the fall, the arc for May is similar to that of July, and the arcs for March and August are similar to each other as well. The red lines indicate the time of day. Where the red and blue lines cross is where the sun will be at any given date and time throughout the year.

The idea of this exercise is to identify each obstacle by the number of degrees above the horizon and the number of degrees east or west of solar (true) south, and then to place a drawing of each obstacle on the diagram.

There are several ways to do this manually. The simplest way is to use a protractor and compass. Tie a string with a weight to the center of the flat protractor base. Set the compass to solar south and look for any obstacles. If there is one, hold the protractor flat side up and look along the protractor to the top of the obstacle. See where the weighted string crosses the degree marker on the protractor and note the number of degrees.

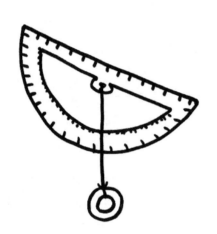

Image 4-6
Image: Kayla Cloonan

In this example, with the protractor presumably pointing at the top of an obstacle like a tree or a building, the string is hanging at 15 degrees. You would then mark the chart at the number of degrees east or west of solar south and 15 degrees above the horizon.

Because of the difference, previously discussed, between magnetic south and true south, called magnetic declination, 180 degrees on the chart is actually 194 degrees (in my part of New Hampshire), so you are looking at true solar south along the flat edge of the protractor, which is 15 degrees above the horizon in this example. On the chart below, this is represented by the X.

After you have completed the siting at solar south, move the compass 15 degrees to the east and repeat the procedure. Then go 30 degrees off solar south and repeat the procedure again. When you have completed your readings to the east of solar south, repeat the procedure to the west until you have completed the chart.

Image 4-7 The view out my window facing south

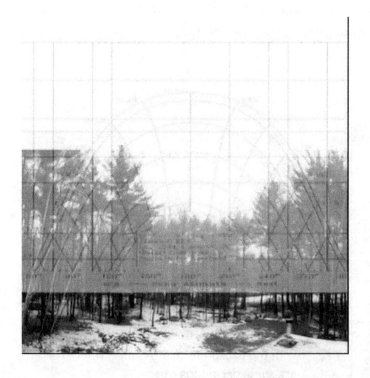

Image 4-8 Through the magic of digital photography, I have superimposed the sun chart on my photo. The line on the chart representing the horizon has been lined up with the actual horizon.

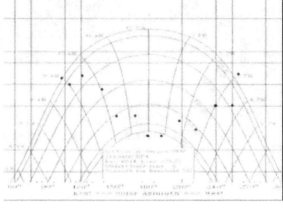

Image 4-9 Using the method described, I have marked the line of obstacles. Notice that in the center I have put dots at the top of deciduous (broad-leafed) trees.

Image 4-10 Here is the resulting set of markers.

Note that these trees will not block the sunlight because by the time the leaves come out, the sun will be higher than the deciduous trees.

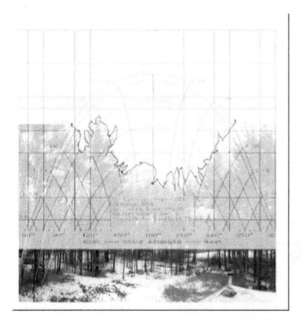

Image 4-11 I've connected the dots. Everything below the black line is shaded, with the exception of the center, where there are some deciduous trees.

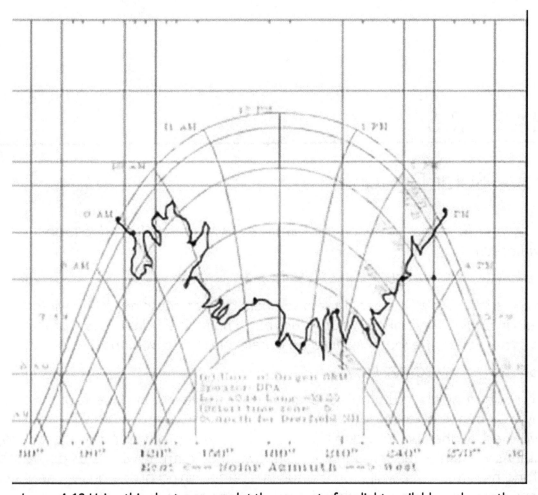

Image 4-12 Using this chart, we can plot the amount of sunlight available each month.

Month	Solar availability	Hours of sunlight/day
December	12-1PM	1
January, Nov.	12-1PM	1
Feb, March, Sept, Oct	10:30-3PM	4.5
April, August	10-3PM	5
May, June, July	9-3PM	6

The specific requirements for different solar technologies vary, but the bottom line for all of them is: the more sun, the better. The solar resource we just evaluated is clearly limited, as one hour a day will provide little benefit. Because I wanted to utilize solar to generate electricity for my house, I had to cut down many trees before I installed a photovoltaic (PV) system. This survey identified exactly which trees I needed to cut down to increase my site's solar exposure.

CHAPTER 5

Passive Solar Design

"Only primitives and barbarians lack knowledge of houses turned to face the Winter sun.

Aeschylus (c. 525/524 BC–c. 456/455 BC)

"Now, supposing a house to have a southern aspect, sunshine during winter will steal in under the verandah, but in summer, when the sun traverses a path right over our heads, the roof will afford an agreeable shade, will it not?"

Socrates (c. 469 BC–399 BC)

You may not realize it, but you have experienced every component of passive solar energy. You can sit by a window with the sun coming in, even if it is below freezing, and feel warmed by the sun. You don't have to do anything but sit passively to enjoy the warmth. As warmth enters through your windows, the warm air will rise as cool air sinks. If you stand next to a brick wall after the sun has been shining on it, you feel heat coming off the wall. These are the principles of passive solar energy.

These principles are well known to my dogs and cat who follow the sun from one window to another throughout the course of each winter's day and who follow the shade all summer. These principles were known to humans centuries ago.

This is a photograph of cliff dwelling ruins in what is now Mesa Verde National Park in southwest Colorado. The cliff dwelling community was built more than two thousand years ago. The cliff you see in this photo faces south. In the winter, the sun's low angle

Image:5-1 Site of Anasazi cliff dwellings in Mesa Verde National Park. This site reflects the major principles of passive solar design.
Photo: Wes Golomb

penetrates deep into the openings where these structures were built. In the summer, the cliff walls shade the whole community from the high angle of the sun. I took this photo on a hot summer's day. It was over 100 degrees on the plateau above this cliff and a comfortable 70 degrees below and inside the cliff.

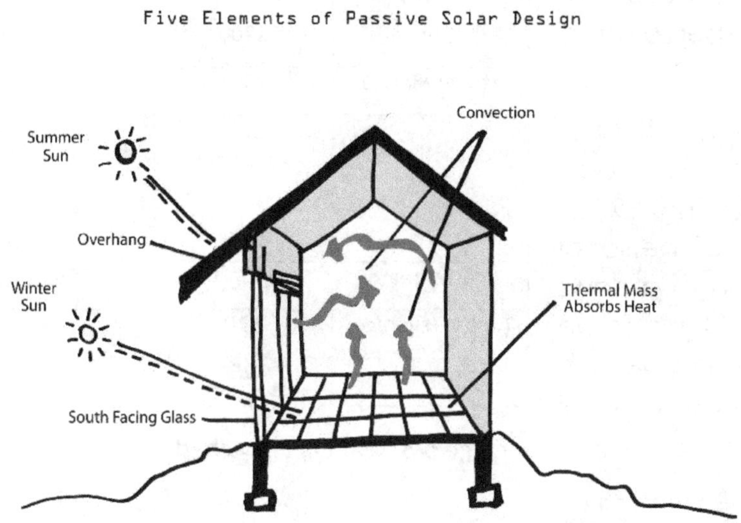

Image 5-2 The Five Elements of Passive Solar Design
Diagram courtesy of U.S. DOE

Passive solar energy is a simple concept. Sunlight falls on a structure, which absorbs the heat, stores it, and delivers it back to the living space when the sun is no longer shining.

The key elements of passive solar energy are:

- South-facing exposure (In the Northern Hemisphere)
- Windows to the south
- Thermal mass (a material such as brick used for heat storage)
- Shade control such as an overhang to limit summer sun
- Heat distribution (a means of moving the heat where it's needed)

Designing a home to take advantage of passive solar is quite simple. You build a well-insulated, air-sealed house with a southern exposure. Place most of the windows on the southern side to maximize the heat gain.

The floor and wall surfaces onto which the sun will shine should be built from a material with a high ability to absorb and store heat. Heat capacity refers to a material's ability to hold heat. Water has a heat capacity of 1. Because of its excellent ability to hold heat, water is the standard by which other substances are measured. Stone, brick, and masonry also have high heat capacities. Wood and metal tend to have lower heat capacities.

The following chart shows the relative heat capacities of various materials. The middle column describes the density of the material in kilograms per cubic meter (kg/m3). The right-hand column describes the heat capacity in Joules (the metric measure of heat energy per cubic meter per degree Celsius).

The Society for the Protection of New Hampshire Forests' headquarters was built with a passive solar design in the 1970s, before the commercialization of personal computers (Image 5-4). The large windows were designed to allow in lots of light, to be stored as heat in a water wall (Images 5-6 and 5-7 below), which provides thermal mass.

Material	Density/kg/m3	Volumetric Heat Cpapacity (J/m3 Deg C)
Water	1000	4186
Concrete	2100	1764
Brick	1700	1360
Stone: Marble	2500	2250
	Materials Not Suited for Thermal Storage	
Plasterboard	950	798
Timber (average)	610	866
Fiberglass	25	25

Image 5-3 Heat capacity chart

**Image 5-4 Society for the Protection of New Hampshire Forests, old wing built in the 1970s.
Photo: Wes Golomb**

Image 5-5 Society for the Protection of New Hampshire Forests, new wing.
Photo: Wes Golomb

The newer addition (Image 5-5), was built after the advent of PCs. Because of the heat given off by personal computers, less external heat was needed to keep the building warm, so the windows are smaller.

Image 5-6 The large windows were designed to allow in lots of light, to be stored as heat in a water wall, which provides thermal mass. Shade provides overheating protection in the hot weather.

Image 5-7 Solar designer Paul Levielle explains passive solar design to my students. The water wall behind him is made from Calwall plastic tubes and is filled with water. When the sun shines on the tubes, the water absorbs heat. When the sun goes down, the water radiates heat back into the well-insulated room.
Photo: Wes Golomb

The above image (5-7) was taken in the early fall. At that time of year, the sun is still high in the sky, leaving the people in this picture in shade. As a result, this well-insulated, air-sealed building stays cool in the summer. As the season progresses and the sun gets lower in the sky, what was once in shadow will be in full sun for much of the day, warming the air and the water tubes.

Several of the houses in this book incorporate some or all of the passive solar principles just described. In particular, the Burns home (Chapter 11) heavily incorporates these principles.

CHAPTER 6

Solar Hot Water

Flat Plate Collectors

Flat plate collectors have been the standard for solar hot water heating for decades. First developed in the 1950s by Hottel and Whillier, they were the most common type of collector installed until the photovoltaic (PV) explosion of the last few years. Flat plate collectors are generally made of copper tubing sandwiched between a heat-absorbing plate and a layer of insulation to retain heat.

The unit is made of an insulated aluminum housing and covered by a protective glass panel, making the collector akin to a compact greenhouse. Inside this box is a heat-absorbing plate, which, depending upon quality, is either painted black or coated in black chrome. A fluid—usually water mixed with a nontoxic antifreeze—flows into the inlet connection of the copper tubing, where it absorbs heat from the collector. The hot fluid then exits the collector via the outlet connection and moves via insulated pipes through a heat exchanger in an insulated hot water tank.

It is important to understand that the water that runs through the collectors is used to transfer heat and is NOT the water that residents drink.

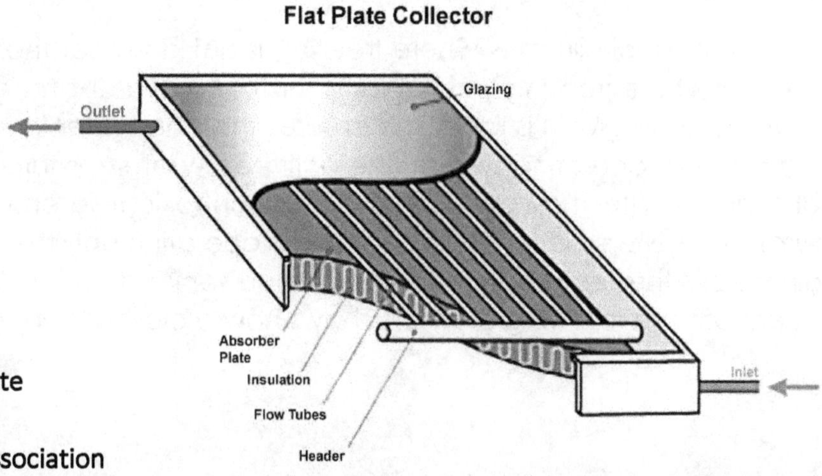

Flat Plate Collector

Image 6-1 Solar thermal flat plate collector
Image: Solar Energy Industry Association

You can see the copper tubes embedded beneath the clear glass and the black absorber plate. The silver reservoir at the top contains the heated water, which is then pumped through a heat exchanger in the bottom of a hot water tank. The heat is thereby transferred and heats the domestic hot water. In warm climates, a storage tank may be mounted above the collectors, allowing the warmer water to rise into the tank in a process called thermosiphoning. This passive solar design can be augmented by a small pump. In cold climates, the storage tank is usually kept inside the building, typically in the basement, where it will not be exposed to cold outdoor winter temperatures. Flat plate collectors are used in both closed loop and drainback systems, each described below.

Image 6-2 Sunda awning evacuated tube collector
Image: Wikipedia Commons

Closed Loop Systems

The closed loop flat plate collector transfers the sun's heat via a fluid that runs through a heat exchanger inside a storage tank and back to the collectors to be reheated. The same liquid is used over and over again. The solar thermal system activates when a temperature sensor in the panel reads 5 degrees warmer than the temperature of the water in the storage tank. Once the system is operating, fluid circulates through the collectors and absorbs the sun's heat. The heated fluid is piped to a water storage tank, usually in the basement. The fluid moves through a heat exchanger—nothing more than a coil of pipe located at the bottom of the storage tank—where the heat is transferred to the water in the tank. The cooled fluid is then circulated back to the collector to be reheated.

In warm or cool climates where freezing is not an issue, the fluid can be water. In cold climates where freezing is possible, a mixture of water and a nontoxic antifreeze called polypropylene glycol is used. This ensures that the water in the collectors will not freeze. In a properly operating system, the antifreeze-water mixture never touches the potable (drinkable) water; however, the use of a non-toxic fluid ensures that humans will not be harmed if there is a leak. The system keeps operating until the temperature in the collector falls below the temperature in the storage tank, at which time the controller turns the system off. The following diagram shows how a closed loop system works.

PRESSURIZED CLOSED LOOP SYSTEM

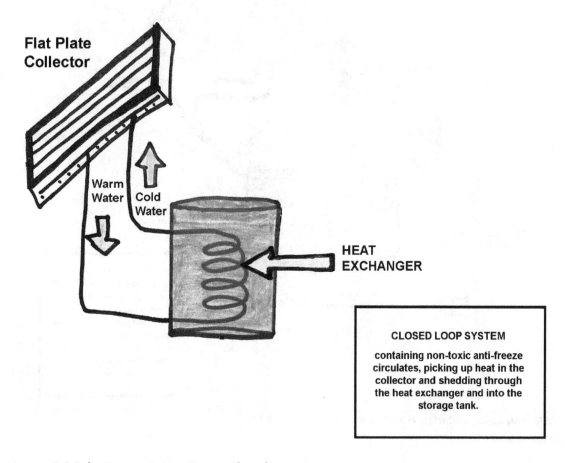

Image 6-3 Solar Domestic Hot Water, Closed Loop System

Drainback Systems

Drainback systems work similarly to closed loop systems but with one difference in the strategy used to protect the fluid from freezing. Instead of using an antifreeze-water mix as the heat transfer mechanism, a drainback system empties all the water out of the collectors and exposed pipes into a separate tank, usually located in a closet on the top floor of the house. When the system activates again, the water is pumped back into the panel. The advantage of a drainback system is that water has a higher heat capacity than antifreeze, making it a better heat-transfer medium. The disadvantage of a drainback system is that its valves, which must open and close, are exposed to the weather, thereby increasing the likelihood of system failure. If the drainback fails to operate and pipes in the collector freeze, a costly repair will be needed.

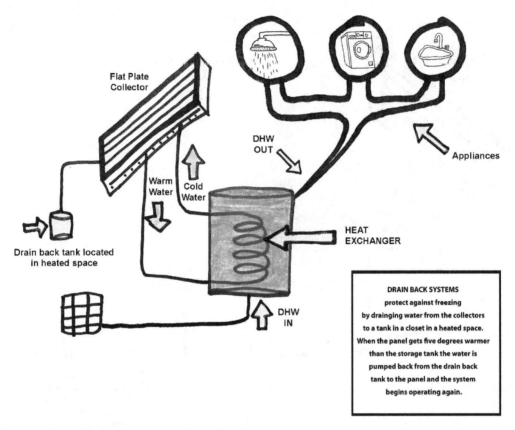

Image 6-4 Solar Domestic Hot Water, Drainback System
Image: Kayla Cloonan

Flat plate collector systems with either antifreeze or drainback freeze protection are run by a controller, which senses the temperature in both the panels and the storage tank. As in the closed loop system, the controller activates the drainback system when the temperature in the panels is approximately 5 degrees warmer than the water in the storage tank. When the temperature in the panels approaches the same temperature as in the tank, the controller deactivates the system. In the case of a drainback system, the water drains out of the panel when the system is turned off.

FLAT PLATE COLLECTOR PERFORMANCE

Some of the factors affecting flat plate collector performance include latitude, season, and solar exposure. Solar thermal systems are less sensitive to partial cloud cover than photovoltaic systems, so a good site will have solar exposure for at least six hours a day, year-round. The amount of hot water a system produces depends on many factors, including the ratio between the surface area of the properly sighted panels and the number of gallons of water to be heated and stored.

A good way to size a domestic hot water system is to start with the amount of hot water consumed by the house's occupants. You can use a flow meter to determine the amount consumed, or make an estimate. Generally, a reasonable amount to plan for is 20 gallons of water per adult and 15 gallons per child per day. In the Northeast and other cool climates, a ratio of 1.5–2 square feet of collector area for each gallon of water to be heated per day is a good rule of thumb. The easiest way to avoid overheating is to size the system for the summer and sacrifice some winter performance. As a result, systems are often sized to produce two-thirds to three-quarters of a family's hot water.

You could size a system here in New England that would provide all the hot water needed for a residence in the winter, but this brings certain challenges with it. Because the sun's radiant energy is weaker in the winter, more collector area is needed to provide energy in cold months. In the winter, solar would provide nearly all of the hot water needed, but in the summer it would provide a lot more than 100% of the hot water needed.

If you've ever used a pressure cooker, you know that water heated in a contained space builds up pressure. If you size your system to provide enough hot water in the winter and don't consider the summer, the system will overheat. Therefore, solar hot water systems must have some kind of pressure relief designed into them.

There are various ways to design a system to deal with this problem. A pressure-relief valve is the most common solution. Back in the 1980s, I sold solar hot water systems for Sears. They used Daystar flat plate collectors. Daystar made panels with a radiator on the top. When the water temperature in the storage tank reached a preset maximum temperature, excess heat would be released via the radiator. From a functional, protective point of view, this is a fine solution. However, there is something troubling about collecting energy and then wasting it by venting it into the air.

I've seen two creative solutions to this problem. In one solar hot water system, sized for the winter, the excess heat generated in the summer is transferred into a large, insulated 2,000-gallon storage tank. The stored heat from the summer is augmented in winter by water that is run through a coil of copper tubing wrapped around the woodstove flue pipe. The coil serves as a heat exchanger, further heating the water in the big tank. The stored hot water is pumped through pipes in the floor as part of a radiant heat space-heating system, which distributes the heat evenly throughout the house.

My favorite solution to the problem of designing a solar hot water system for maximum winter gain was in a house I saw some years ago. The owner had run an extra line from the system to his side yard. There he transferred the excess heat into pipes running through a concrete slab in his backyard, à la radiant heat. He stacked his green cordwood on the slab, and the heat radiating off the slab dried his cordwood in a fraction of the time it normally takes to season wood.

Evacuated Tubes

While flat plate collectors have been the standard for solar hot water for years, there is a newer system that has gained popularity. Evacuated tubes collect the sun's energy in a different way. These collectors are glass vacuum tubes with a copper heat-absorber chamber inside. The vacuum allows the sun's shortwave radiation to enter the tube and heat the absorber chamber, while letting far less long-wave radiation out. The sun's radiation heats a fluid, usually a mixture of alcohol and water, inside the absorber chamber. The fluid and its contained heat rise to the top of the absorber chamber, which is fitted into a manifold. The heat is then transferred from the tube to water (or a water-antifreeze mix), where it flows to the storage tank.

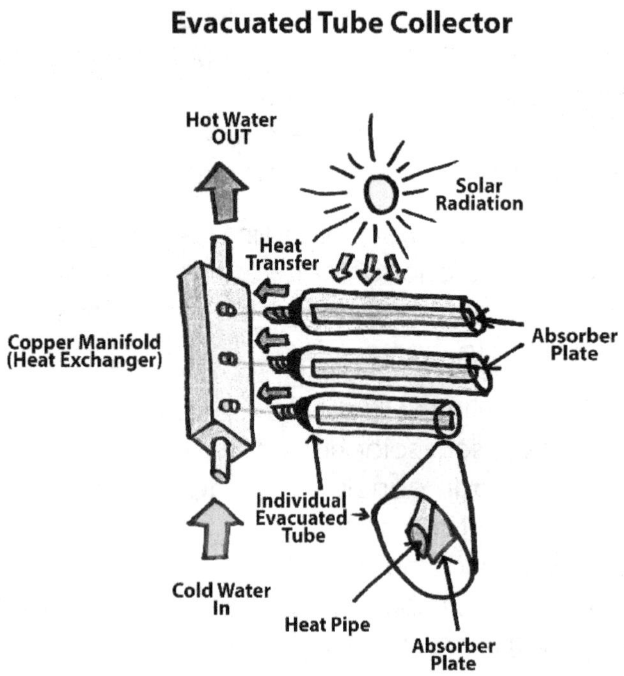

Image 6-5 Evacuated Tube Collector
Diagram: Kayla Cloonan

Image 6-6 Evacuated Tube Ground Mount
Photo courtesy Ra Boe and Wikipedia of
Wikimedia Commons

FLAT PLATE vs EVACUATED TUBE COLLECTORS		
Factor	Flat Plate	Evacuated Tube
Relative cost	Generally less expensive	About 15% more expensive
Cost per btu	Generally less expensive	
Warranty	Differs by manufacturer; average 20 years	Differs by manufacturer; average 10 years
Installation	Heavier and more awkward	Lighter but more fragile
Highest temperatures	180°F	250°F Possible overheating hazard
Cold-temperature performance		Performs better in cold weather
Appropriate usage	Domestic Hot Water (DHW;) space heating	Higher temperature commercial use; space heating
Performance in snow	Sheds snow quicker	Vacuum acts as insulation; slower melting
Structural wind loading		Lighter and allows wind to pass through

Opposite page: Image 6-7 Evacuated tube Awning Mount
Image: Pablo Fleishman, Geosolar Store, Keene NH

Either one of these systems, if designed and installed well, will work effectively for a long time. In 2008, more than twenty-five years after I sold solar energy for Sears, I met someone to whom I'd sold a flat plate collector system. He told me it was still fully operational, though he'd just replaced the water tank. He had a horse farm and estimated that the system saved him from heating somewhere around 1,000 gallons of hot water a year. That's roughly 25,000 gallons in the twenty-five years since he first installed the system.

CHAPTER 7

Solar Electric (Photovoltaic) Energy

In 1955, the Hoffman Electronics Semiconductor Division created a 2% efficient commercial solar cell. The cost was $25 per cell, or $1,785 per watt. This technology made the space program possible because it filled the growing need for reliable, continued electricity to power satellites.

In 1958, Hoffman Electronics created 9% efficient solar cells, which were used on Vanguard I, the first solar-powered satellite. Vanguard I had a 0.1 W, 100 cm² solar module. Solar panels also powered Telstar, the first communications satellite to broadcast live TV between North America and Europe. The satellite's fame was enhanced by the 1962 hit record of the same name by the Tornados (cue the music here...). But at these 'astronomical' costs, solar photovoltaics were little more than a curiosity.

As you can see in the graph below, the price of photovoltaic cells has continued to fall. I owned a solar business in the early 1980s. The lion's share of our business was passive solar additions and domestic solar hot water systems. The only PV system I sold was to a man from New Hampshire who was a missionary home on a visit from Africa. He wanted a small PV system so he could power a refrigerator in a region that had no other electricity for hundreds of miles. The distributor we bought his components from was owned by Don Kent, Boston weatherman and early solar proponent. Things have certainly changed since that time.

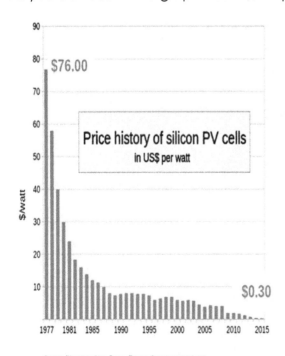

Image 7-1: Price history of silicon PV cells.
Image: Wikipedia Commons

In the past five years, the cost of making energy from the sun has dropped dramatically. In 2020, installing solar on a new house was less expensive than paying electric bills. This has made it much more realistic and cost-effective to build and live in a net-zero home.

The solar revolution has been a long time coming, but there has never been a better time to go solar than now. Solar's potential can be seen in the New York Times headline below, from an article announcing that German PV was generating more energy over a couple of days than was being used. The utility was actually paying people to use power. Look at the chart comparing German solar potential with that of the United States. Most of Germany has potential similar to Alaska. If they can do it, so can we.

Power Prices Go Negative in Germany, a Positive for Energy Users[10]

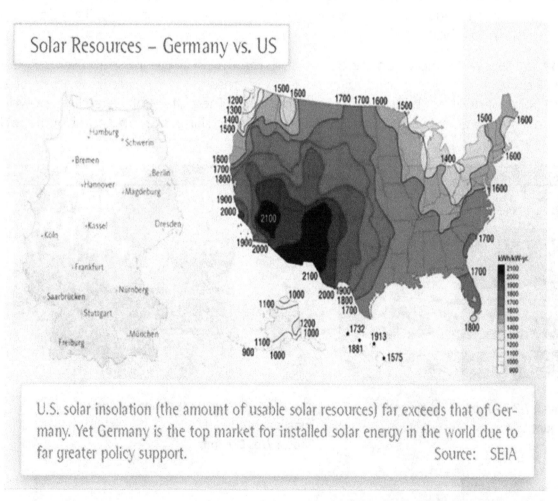

Image 7-2 Comparison of German and US Solar Resources

There has never been a better time than now to install solar. If you have a sunny location, the rate of return is usually better than the stock market.

Electricity from the Sun

The process of converting sunlight to energy is carried out by individual photovoltaic cells. The modern PV cell is made of silicon. When sunlight excites the surface, the electrons with their excited energy move through the circuit, where they give up their energy to do the intended work in the form of electrical power. When sunlight falls on the cell, a direct current is generated. The cells are wired together and arranged into modules, which are arranged into an array (sometimes referred to as a solar panel). A module produces DC power, which is then converted to AC current by an inverter. Until recently, an array had only one or two inverters, and if one module became shaded or failed, it would bring the whole system to a halt.

In the last few years, individual modules have come with micro-inverters that convert DC to AC current for each individual module. As a result, if a module is shaded or stops functioning, it does not bring down the whole array. The potential generating power of a system is measured in kilowatts and depends upon the siting, size, and number of modules in the array.

Image 7-3 PV module on ground mount

Image 7-4 Worker wires a module during installation
Photo: Wes Golomb

Image 7-5 Inverters support an 18-module, 2-array system.

Image 7-6 Ground mount of 18 modules, 2 arrays.

Image 7-7 3.2 kW fixed ground mount PV system at Prescott Farm Environmental Education Center This array at Prescott Farm was installed by students, mostly electricians, being trained through a North American Board of Certified Energy Practitioners (NABCEP) certificate program. The class trains electricians to install photovoltaics.
Image: Wes Golomb

We know the sun moves across the sky by day, and the arc it takes changes over the course of a year. We also know that the most radiation is collected when the sun's rays strike the module at a 90-degree angle. Because the module is fixed, whatever angle we mount it at will only be ideal a small percentage of the time. Whichever angle we choose will be a compromise. The amount of radiation that can be collected depends upon the angle of the sun to the collector, as well as the direction the collector faces (solar south being ideal). The accepted compromise is usually an angle equal to the latitude of the project. My house in New Hampshire is at a latitude of 44 degrees north, so the ideal angle for a fixed mount system facing toward solar south is 44 degrees. As you move the collectors away from that ideal angle, the performance drops off.

There is an alternative to the fixed module compromise: a 2-axis tracking system as seen below.

The tracking system mounted in the foreground follows the sun and keeps a right angle to the sun's radiation. This maximizes the solar gain and increases the efficiency of the system by as much as 40%. The system uses a motor controlled by a computer to keep the modules at a 90-degree angle to the sun. The system on the roof, in the background, faces almost due south. In this photo, taken in the morning, you can see that the tracking system is facing more toward the east than the fixed roof mount is. As the sun moves, the tracking system modules follow it, keeping the angle of the modules at or close to the optimum 90 degrees to the incoming solar radiation. The tracking system collects as much as 45% more energy than the roof mount system, some of which is used to move the

Image 7-8 Tracking system in foreground, roof mounted system in background
Photo Wes Golomb

tracker. Tracking systems mean moving parts exposed to weather. This may also mean increased maintenance expenses, though I have yet to hear of significant problems with tracking systems.

Storing PV Energy

PV cells convert sunlight to electricity, which is great when the sun is shining. What happens when the sun does not shine? A way to store the electricity is needed.

A network of batteries can be installed to store surplus power. Battery storage offers backup power when the sun does not shine or there is a power outage. Until recently, deep-cycle lead-acid batteries were the standard. These batteries require proper ventilation, regular maintenance, and periodic replacement. The laws of thermodynamics tell us that when we convert energy from one form to another—in this case from electricity (kinetic energy) to battery storage (potential energy) and back again—there must be some energy loss to the system. As a result, storing collected solar energy in batteries has been relatively inefficient. Recent advances in battery technology are beginning to change that.

Lithium-ion batteries have been commercialized and are now the most-used form of storage for residential applications. And their use is growing rapidly! According to the International Energy Agency, installation of residential battery storage doubled in 2019.

Image 7-9 Thirty 9-volt batteries store energy for this off-the-grid home. Note the ventilation on either side of the box. Thanks to Chris Pinkham.
Photo: Wes Golomb

The best-known battery storage option is the Tesla Powerwall. It increases the user's personal consumption of the solar energy produced by storing it directly rather than putting it back on the grid. A battery backup can be used in on-grid and off-the-grid situations. Batteries typically do not power a whole house but take the place of a small generator when the power goes out. In service areas where there are different electric rates charged at different times of day, batteries allow the user to use solar energy at peak periods when utility rates are highest. Battery storage can work in tandem with net metering or be used in off-grid situations.

Tesla estimated the 2020 cost of Powerwall hardware and installation to be between $9,500 and $10,400. The Economics of a Tesla Powerwall 2[11], a Stanford University study done in 2016, concluded that if you pay more than 0.139 cents/kWh for electricity, the use of the Powerwall 2 will at least pay for itself over the course of ten years. This study is a bit out of date, because in the past four years, the costs for battery storage have come down and utility rates have risen.

Net Metering

Until recently, if you wanted to use solar-generated electricity at night, batteries were necessary. An alternative, net metering, has become popular in the last decade. When a system is net metered, any surplus (unused) electricity that is produced is fed back onto the utility's power grid through the electric meter. The meter actually runs backward, measuring the electricity produced as well as consumed. The producer of the net-metered power gets a credit, and when there is not enough sun to meet the system's demands, the needed energy is once again drawn from the grid, running the meter forward. This isn't really storing the energy. It's more like depositing and withdrawing energy, the same way you don't get the exact dollar bill that you put into your bank account back when you withdraw it. The energy you put back on the grid is used by your neighbors.

In addition, a number of states have adopted group net metering, which is essentially an accounting device allowing the distribution of electricity, all generated from the energy stored by multiple people on one system, to small communities of people via the grid.

Photovoltaic technology is scalable, meaning that it can be used for houses, neighborhoods, communities, or commercial buildings, or centralized at a utility scale. I don't have enough sun for PV in my backyard. My neighbor, however, has a big field with no trees in his backyard. Suppose he installed enough PV to supply both of us? He could put his excess onto the grid, and I could use it. This sort of arrangement can be formalized in places that allow group net metering.

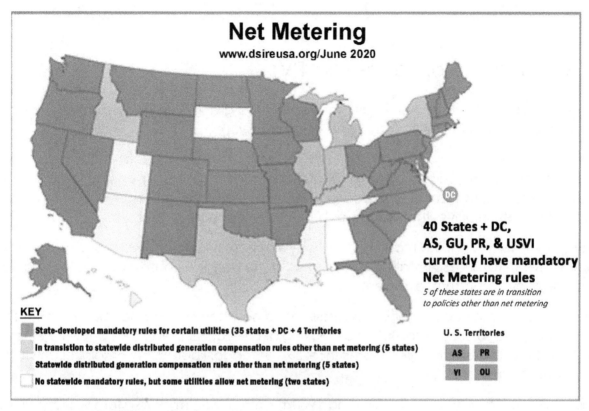

Image 7-10: US States that have Net Metering programs
Photo courtesy of dsireusa.org

One shortcoming of net metering is that when the power goes out on the grid, the net-metered power necessarily goes out too. During a power outage, there can be no unknown systems dumping power onto the line when workers are trying to fix it, so there is a cutoff switch that engages when the power goes down. It is possible to wire individual backup battery systems, which can be used when the grid is down. This, however, requires the additional expense of installing and maintaining batteries.

Overall, net metering makes the grid more stable. For the last twenty years, as the climate has warmed and more people have started using air conditioning, the peak periods of power use have changed from winter to summer. Fortunately, this is usually the exact same time—hot, sunny days—when solar panels are generating large amounts of electricity, more than their owners need, so lots of unused solar energy is put back onto the grid. This avoids brownouts and blackouts. It also may replace the need to start up secondary coal-fired power generation, like New Hampshire's dirty, expensive Merrimack Station in Bow.

In this way, having net-metered solar generation makes the whole grid more resilient.

The Economics of Residential PV

Photovoltaics is the proverbial "gift that keeps on giving." A PV system will typically provide a home with thirty to forty years of electricity.

PV is one of the last major items to be installed when a new home is built. The expense at this time seems daunting to many homeowners who are already spending huge sums of money, often more than they expected to. There is an urge among many to put it off. Though counterintuitive, that might not be the best economic decision.

With proper financing, the monthly cost of a PV system is less than the cost of purchasing electricity off the grid, so installing a PV system at the time the house is built actually lowers short-term and long-term expenses.

The difference may be harder to evaluate with a new home because there is no electric bill to compare solar costs to. But when looking at a retrofit, the numbers become clearer. Let's consider my retrofit in New Hampshire. Six years ago, I put 6.5 kW of solar on my thirty-five-year-old house.

I used to have an electric bill of around $200/month. When my PV system was installed, I got two loans. The first was a short-term note for the amount of the federal tax credit and the New Hampshire solar incentive, which I paid off that year. The second was a twelve-year loan that costs $125/month. I am now getting between 85% and 90% of my yearly electricity from solar. (I was limited in roof size and unshaded areas.) My electric bill, mostly a service charge now, is never more than $75. Since installing my system six years ago, I've paid less for my residential energy EVERY MONTH.

When my loan is paid off in six more years, my loan payments will have been less than or equal to what I would have paid the electric company every month for electricity made from nuclear and/or fossil fuels. After the loan is paid off, the system should give me another twenty to twenty-five years of free electricity.

The above logic and financing is based on a real comparison with previous electric bills. It is harder to do this calculus when the house is new and there is no means to make direct comparisons. Given the appropriate financing and a good site, I believe the logic and economics of the above example carry over to the question of when to install solar on a newly built home. The answer is NOW!

Here's an example breakdown of a standard rooftop-mount 7-kilowatt solar electric array (twenty-two 320-watt panels), taken from a local solar installer's website in July 2020[12]. This system will produce around 8,450 kilowatt-hours of electricity per year, enough to meet 100% of the needs of many moderately sized single-family homes.[13]

- $22,880 Gross installed cost
- ($6,864) 26% federal tax credit
- ($1,000) NH state rebate; check http://dsireusa.org/ to see if your state has incentives.
- $15,016 total investment

Say you pay 16 cents per kilowatt-hour now for electricity. By saving you around $1,344 per year, the PV system will have paid for its installation cost in roughly nine years and will continue producing free power for the next twenty-plus years. Solar power equipment is warrantied for twenty-five years, and we expect it to last for more like forty-plus years.

This example is from a reputable solar dealer in New Hampshire. I use it only as an example. Every site is different, and every family's lifestyle and demand for electricity is different.

One of the shared hopes of those who study and advocate for PV is that efficiency will continue to improve. There are several technologies on the horizon which will likely improve efficiency and/or lower costs.

Thin-film PV cells are made from extremely thin layers built up on a substrate such as glass, polymer, or metal. This technology requires about 1% of the materials needed for first-generation cells and can be produced using efficient manufacturing processes. Other promising technologies include dye-sensitized solar cells, organic solar cells, and concentrating photovoltaics.

There will continue to be technological improvements that will probably come incrementally, and either the upfront cost will come down or the value for the price will go up.

The cost of installing a PV system, particularly in a new home, is less than the cost of running wires and getting electricity from your local utility.

Don't wait—do it now!

CHAPTER 8

Air- and Ground-Source Heat Pumps

Put a glass of 50-degree water into the refrigerator for an hour. Come back and you'll find the water at 40 degrees. We know that energy cannot be created or destroyed, so what happened? Where did that heat energy go? The refrigeration cycle has removed heat from the refrigerator and directed it out the back, where the dog hair collects. In the process, your water is cooled. Instead of wasting energy out the back of the fridge, suppose that heat was collected and concentrated for use?

Both ground-source geothermal and air-source heat pumps use the refrigeration cycle. In cold weather, these heat pumps remove heat from outdoor air or groundwater and move it into the home, where it is distributed throughout the cooler building. In hot weather, the process is reversed and heat is moved from the building to the cooler outdoor air or groundwater.

We know from basic laws of thermodynamics that heat naturally flows from warm to cold environments, so at first glance it is totally illogical to be talking about using 20-degree air or 50-degree groundwater to heat the inside of a house to 70 degrees. However, the refrigeration cycle makes it possible to do so in a very efficient manner.

There are a few simple concepts behind the refrigeration cycle:

> 1. When a liquid absorbs enough heat, it boils (evaporates) and changes state from a liquid to a vapor. An example is water in a tea kettle absorbing a lot of heat energy, boiling, and turning into steam.

> Vapor (steam) contains the heat that it absorbed when it evaporated from a liquid. When vapor condenses, changing back to a liquid state, it releases that heat.

> 2. The boiling temperature of any liquid can be raised by raising the pressure of the liquid. A pressure cooker is a practical example of this. The pressure raises the temperature inside and cooks the food faster than a pot with a lid set on top. Conversely, the boiling temperature of any liquid can be lowered by lowering the pressure of the liquid.

3. The condensing temperature of any liquid can be lowered by increasing the pressure of the vapor. (Think of this as wringing out a sponge.)

A heat pump is used to transfer heat from a source of heat energy (such as outside air or groundwater) to a thermal reservoir, which in the case of a heat pump is the air inside the house, and in the case of a geothermal pump is usually a water storage tank in the basement. The components in a heat pump include: two heat exchangers, an evaporator, an expansion valve (which changes the pressure), a compressor, and a chemical called a refrigerant (like Freon), which is a gas at room temperature and a liquid when cooled or compressed.

Image 8-1 Refrigeration cycle #1
Diagram: Kayla Cloonan

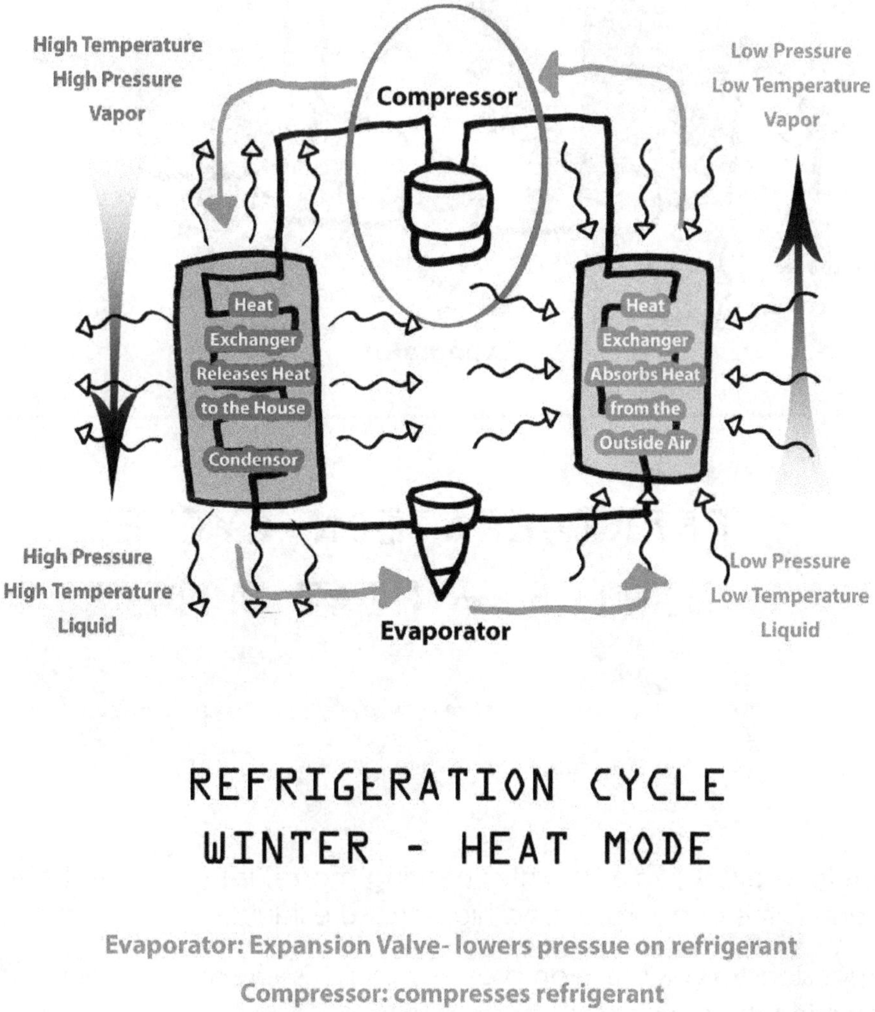

REFRIGERATION CYCLE
WINTER - HEAT MODE

Evaporator: Expansion Valve- lowers pressue on refrigerant

Compressor: compresses refrigerant

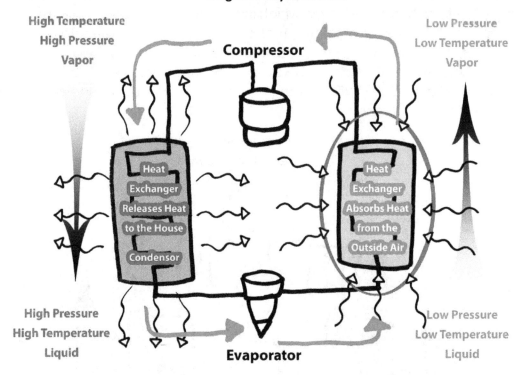

Image 8-2 Refrigeration cycle #2
Diagram: Kayla Cloonan

REFRIGERATION CYCLE
WINTER - HEAT MODE

Evaporator: Expansion Valve- lowers pressue on refrigerant

Compressor: compresses refrigerant

Looking at the refrigeration cycle in winter heating mode, let's start at the evaporator. The refrigerant enters the evaporator as a high-pressure liquid.

Through an expansion valve, the high-pressure liquid is expanded, thus reducing the pressure and lowering the refrigerant's boiling point.

The refrigerant then flows to the heat exchanger, where it absorbs heat. from the ground which evaporates (turns into a gas) the refrigerant. At this stage, the refrigerant has absorbed a lot of heat in the same way a sponge absorbs a bunch of water.

Once all of the refrigerant has evaporated into a vapor, it continues on, still carrying all the heat. At this point, the system will feel cold to the touch, but in reality it contains a lot of heat.

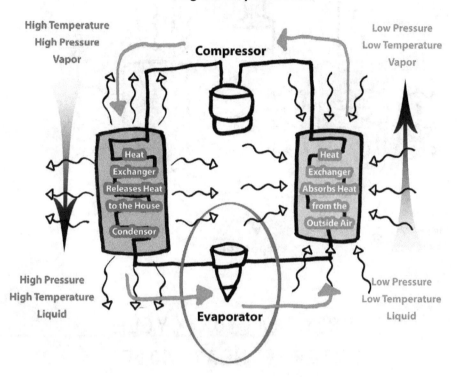

Image 8-3 Refrigeration cycle #3
Diagram: Kayla Cloonan

REFRIGERATION CYCLE
WINTER - HEAT MODE

Evaporator: Expansion Valve- lowers pressue on refrigerant

Compressor: compresses refrigerant

The compressor then moves the refrigerant to a heat exchanger. The compressor pumps the vapor refrigerant along while increasing the pressure of the vapor. As the pressure increases, it forces the molecules closer together and reduces the condensing temperature of the vapor.

Image 8-4 Refrigeration cycle #4
Diagram: Kayla Cloonan

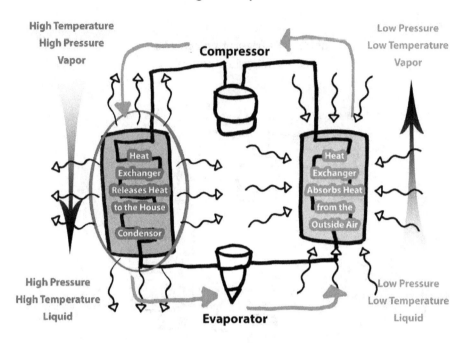

REFRIGERATION CYCLE
WINTER - HEAT MODE

Evaporator: Expansion Valve- lowers pressue on refrigerant

Compressor: compresses refrigerant

The pressurized vapor next enters the condenser and is condensed back into a liquid. When this happens, the vapor releases all of the heat it absorbed in the evaporator. The condenser will feel very hot because the vapor is under pressure and releasing the absorbed heat. This is like wringing out the sponge that absorbed water.

The condenser acts like the radiator in a car. The air that passes through removes the heat from the refrigerant. During winter heating mode, the condenser is releasing all the heat it absorbed into the interior of the home. The now cooled refrigerant is condensed turning all the vapor back into a liquid that has very little heat in it.

The compressor continues pumping the liquid back to the evaporator as a high-pressure liquid and the process starts all over again.

During summer cooling mode, the refrigerant flow is reversed, which means the coil in the home becomes the evaporator and the heat exchanger outside the home becomes the condenser. Heat will then be moved from the inside of the home to the outside.

In short, the refrigeration cycle moves existing heat from one location to another simply by using pressure to force the evaporation and condensation of a liquid. The electrical energy consumed by the equipment is only used to move the refrigerant, change its pressure, and operate some fans and/or pumps. The energy consumed is not used to produce heat (thermal energy) but only to move it. This is where the efficiency gains of heat pumps come from when compared to other heating methods that convert fuel to thermal energy.

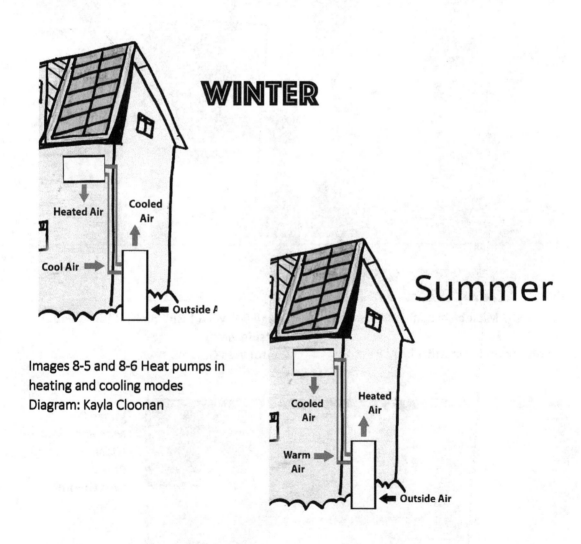

Images 8-5 and 8-6 Heat pumps in heating and cooling modes
Diagram: Kayla Cloonan

Coefficient of Performance (COP)

The coefficient of performance (COP) is a measure of a heat pump's efficiency. It represents the ratio of energy output to energy input. Typically, 1 unit of input energy provides 2 to 5.5 units of output heat energy. Upcoming chapters will highlight several houses that use heat pump technology; geothermal and mini-split air-source heat pumps.

Image 8-7 Mitsubishi mini-split (outside view)
Photo: Courtesy of Mitsubishi Electric

Image 8-8 Mitsubishi mini-split installed (outside view)
Photo: Wes Golomb

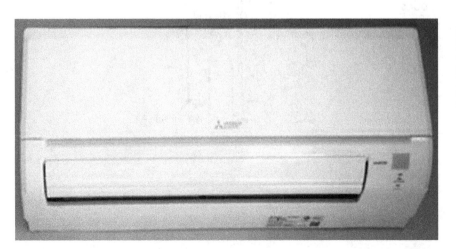

Image 8-9 Mitsubishi wall unit (close-up)
Photo: Wes Golomb

We will also examine a house that uses heat pumps at higher temperatures in a greenhouse to supply heat for a home. In this case, because of the high starting temperatures of the greenhouse, the COP is around 8.

Solar and/or wind energy makes these systems even more efficient by generating the needed operating energy without using fossil fuels. This extra leverage is a key strategy for heating and cooling net-zero homes.

Air-Source Heat Pumps

Air-source heat pumps are a very efficient way to heat and cool a home. In the winter, they use the refrigeration cycle to extract heat from the outside air and move it into the home via a refrigerant. A modern, high-performance system can provide full heating capacity down to 5°F and will provide heat to below -10°F at a progressively lower efficiency.

Because of heat pumps' efficiency, these units are very popular in areas with cold climates. More than 50% of the houses in Norway have heat pumps installed. According to the Mitsubishi Electric Corporation, more than thirty thousand heat pumps have been installed in Maine since 2010. Thousands were installed north of Bangor with few issues or complaints. They offer ongoing reliability and efficiency even at the coldest temperatures, which fall well below 0.

Image 8-10 Mitsubishi wall unit
Photo: Wes Golomb

Image 8-11 Mitsubishi ceiling module; replaces ducted system
Photo: Courtesy of Mitsubishi Electric

Ground-Source Heat Pumps (Low-Temperature Geothermal Energy)

Because of the constant influx of heat from the sun, the groundwater temperature at 25 feet below the surface is near 50°F year-round from Caribou, Maine, to Manhattan, New York. Once you drill below 500 feet, there is a gain of 0.75°F for every 100 feet further down. In Manhattan, a 2,000-foot bore hole for a skyscraper yielded 60°F water. Ground-source systems use the refrigeration cycle to remove heat from the ground or groundwater and provide heating and cooling for a home.

Earth Coupling Systems

In addition to the heat pump, a low-temperature geothermal energy (LTGE) system requires a means for moving heat from the earth into the system. This is accomplished through a series of pipes, known as a coupling system, that are installed in the ground to access the heat. There are a variety of pipe configurations that can be used to bring heat to the heat exchanger. These coupling systems fall into two general categories. Closed loop systems use water as a heat-transfer mechanism, with water recirculating through a loop, picking up heat from the ground and carrying it to the home where the heat is extracted. Closed loop systems may be vertical or horizontal. Open loop systems use groundwater directly for their heat source.

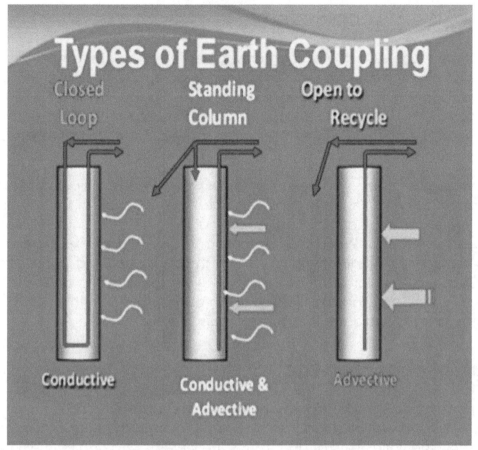

Image 8-12 Types of earth couplings
Diagram: Courtesy of Martin Orio, Massachusetts Geothermal Services

Vertical Closed Loop Systems

A vertical closed loop system works by cycling a fixed amount of water through a closed loop of pipe. Cool water flows down into the well. Conduction from the groundwater warms the cool water as it circulates. That water then flows back to a heat exchanger, which extracts the heat that has been transferred from the well.

These systems are relatively expensive because of the cost of drilling wells. However, because wells take up little surface area, this configuration can be used almost anywhere with the proper geological profile. Often a well for domestic water can also be used for geothermal systems, lowering the cost significantly.

Horizontal Closed Loop Systems

A horizontal closed loop system uses the constant temperature of the earth below the frost line to supply the home with heat in winter and to dump excess heat in the summer. For this configuration to work, trenches are dug six feet deep and pipe is laid in the trenches. In contrast to vertical closed loop systems, horizontal installation requires a lot of surface area and must be free of rock ledge near the surface. Horizontal closed loop systems are typically less expensive.

Horizontal closed loop systems work well in locations where water can percolate through sandy soil back into the same water table it came from. This type of well is popular in areas where there are terminal moraines. A terminal moraine is the southernmost point where a glacier advanced. When the ice melted, it dropped all the material it had pushed and carried with it as it moved south over the centuries. Long Island, New York, and Cape Cod and the islands off Massachusetts are terminal moraines. Non-sandy inland areas, which cannot absorb water back into the ground at the rate it is being returned, are not ideal candidates for horizontal closed loops.

Standing Column Open Loop Systems

The standing column well is a hybrid system. An open-ended pipe takes water directly from the bottom of the well at around 50°F, pumps it through the heat exchanger, and returns it to below the water level at about 40°F. The well is recharged by the 50°F water table over time. If recharging cannot keep the well temperature near 50°F, a standing column well can be bled. To do this, the temperature of the well is increased by pumping some of the colder water into a dry well for short periods of time. This colder water is then reheated by the earth as it makes its way back to the water table. Because the well now contains less 40°F water, the warmer recharge water can more quickly heat the colder well water.

In the summer, the system runs in reverse. If the well temperature gets too warm, the same bleeding process can be used to cool the well off to maintain efficiency. This is an important factor to consider because the heat pumps become progressively less efficient as the source temperature falls (in winter) or rises (in summer) beyond the temperatures that yield the highest efficiency.

There should never be free-falling water. The return pipe to the well should be below the water level. When free-falling water becomes oxygenated, it combines with iron, providing a food source for a strain of bacteria that metabolizes iron. The standing column well should also be electrically grounded. This avoids electrolysis, which will otherwise corrode the metal in the system, particularly in areas with salty water.

Both ground-source geothermal systems and air-source heat pumps can leverage other sustainable technologies. These technologies run on electricity, and one unit of solar- or wind-generated electricity input into a geothermal heat system can yield four or five units of heat energy. For this reason, many zero-energy buildings use both a renewable electricity source and ground-source geothermal or heat pumps.

Comparing Air-Source and Ground-Source Heat Pumps

As we've seen, both types of heat pumps use the refrigeration cycle. Each has its own advantages and disadvantages. An air-source heat pump is less expensive to install because there is no well to drill.

However, because the groundwater in most areas stays in the 50°F range, and air temperatures can fall well below freezing, ground-source systems are a bit more efficient because there is less "lift" needed by the system.

A recent Massachusetts study[14] for Zone 5 (New England and equivalent temperatures) compared the annual coefficient of performance (COP) for heat pumps and found that the annualized COP for air-source heat pumps was 2.5 at 45°F. As the temperature dropped, so did the efficiency.

The same study found that the annualized COP for ground-source systems was between 3.5 and 3.7 depending on configuration and geography.

There are two basic reasons for this difference. To heat a house to 70°F, ground-source systems only need to raise the temperature by around 20°F, as opposed to air-source heat pumps, which may have to raise the temperature by 70°F or more to reach the desired temperature.

Further, consider the heat-transfer materials. Water is about four times as conductive as air. You can put your hand into an oven at 212°F for a few seconds. Don't try that in water! The difference you would feel is due to water's higher conductivity.

Typically, the lifespan of a ground-source geothermal system is longer than that of an air-source heat pump because the components are either below ground or inside the house, whereas the air-source heat pump has an outside unit that is constantly exposed to the weather.

On the other hand, air-source heat pumps take up less space than geothermal systems because there is no need to drill a well.

Both systems are currently eligible for a federal tax credit, and many states have further incentives to install these technologies. Several years ago, the U.S. Department of Energy described ground-source geothermal systems as one of the best energy-saving investments available.

Image 9-1 The Wallace-Brill home
Photo: Wes Golomb

CHAPTER 9

The Wallace-Brill Home

John Wallace and Lessa Brill raised their children in the rural town of Barrington, New Hampshire. After their kids fled the nest, John and Lessa wanted to downsize to an energy-efficient home. Their goal was to find a place where they could live in retirement for the rest of their lives. One of their key requirements was a home not heated primarily with wood. As John put it, "I'm not getting any younger and I don't want to have to cut and split wood for my primary source of fuel."

Barrington had been their home for the previous twenty-plus years, and they wanted to stay there, but they could not find a house that was both energy-efficient and to their liking. So they decided to build a new house instead of purchasing an existing one. They found and purchased some very nice land within sight of a pond and began looking for a builder.

They met Bob "The Builder" Irving at a workshop where he was speaking about building net-zero ready homes. One of the things that stuck with them was Bob's idea that "Any style house can be net-zero if built correctly." They were also pleased with his willingness to design the house with them rather than hiring an architect. And what put them over the top was hearing Bob's company sponsoring the local NPR station.

The home they built has similarities to Mike Marion's home described in Chapter 3. There are double walls set ten inches apart with blown-in cellulose in the space between them.

Images 9-2 and 9-3 Double walls set on slab on grade
Photo: Wes Golomb

Images 9-4 and 9-5 Triple-pane windows are carefully installed, caulked, and taped so they are air sealed.
Photos: Wes Golomb

There are also some significant differences in this home. The most obvious difference is that the Wallace-Brill home has no basement. Instead, it is built on a slab on grade, with traditional post-and-beam construction.

Image 9-6 Slab on grade construction
Photo: Wes Golomb

Image 9-7 Post and beam (notice the pegs)
Photo: Wes Golomb

The house they designed is mostly one-story, with the bedrooms, bath, kitchen, and living room on the first floor. They built a single room on the second floor—an observatory, which, in my humble opinion, is one of the coolest rooms I've seen while profiling sustainable homes. The observatory looks out over the pond and woods and gives one a sense of peace and tranquility.

Image 9-8 Observatory and deck
Photo: Wes Golomb

The process of building was, in fact, not peaceful and tranquil. Constant rain and issues with backorders of materials made this build a much longer and more involved process than was expected.

Image 9-9 Inclement weather and delays required a tarp covering until the roof could be completed. Photo: Wes Golomb

The construction of the roof also presented an interesting challenge. It needed to be air sealed and super insulated and was, of course, a design that limited thermal bridging. The owners preferred not to use foam as an insulator. The challenge was to satisfy all these requirements and produce a finished ceiling of 2x6 spruce.

The spruce sits right on top of the beams, so the builders had to build from the bottom up. A peel-and-stick membrane was applied to the spruce ceiling, above which sit 2x6 TJIs (engineered floor joists & rafters made from wood chips). This met all the structural requirements.

The TJI bays were filled with R-60 batts of rock wool insulation, which was covered with half-inch OSB sheathing. The roof is strapped with 2x4s above TJIs, allowing the needed air space to vent any moisture so as to keep it from accumulating. The roof assembly was then covered with half-inch AdvanTech, a type of high-density sheathing that looks like an oriented strand board but is water-resistant. The AdvanTech was covered by a roofing membrane fabric, and lastly, a low-maintenance standing seam metal roof was installed.

Image 9-10 Post and beam with 2x6 spruce ceiling
Photo: Wes Golomb

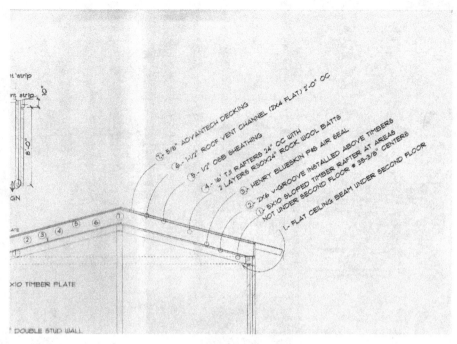

Image 9-11 Roof construction detail
Drawing courtesy of Robert Irving, RH Irving Homebuilders

A beautiful, unique spiral staircase, built by John and Lessa's nephew, leads to the observatory.

Image 9-12 Spiral staircase
Photo: Wes Golomb

Image 9-13 Spiral staircase
Photo: Wes Golomb

Other Energy-Efficient Features

Mechanical Ventilation

Like all tight houses, the Wallace-Brill home needs mechanical ventilation. Instead of using the heat recovery ventilation (HRV) system described in Chapter 3, this house uses a Lunos ventilation system.

Image 9-14 Lunos ventilation unit (one of a pair)
Photo: Wes Golomb

The Lunos ventilation system utilizes pairs of vents installed through the wall units, as pictured above. According to the company, the units operate continuously but can be turned off with a switch. They are installed in synchronized pairs for balanced ventilation. One unit supplies air, the other exhausts air, and they switch direction at one-minute intervals.

**Images 9-15 Lunos ventilation path diagrams.
Courtesy of 475 High Performance Building Supply**

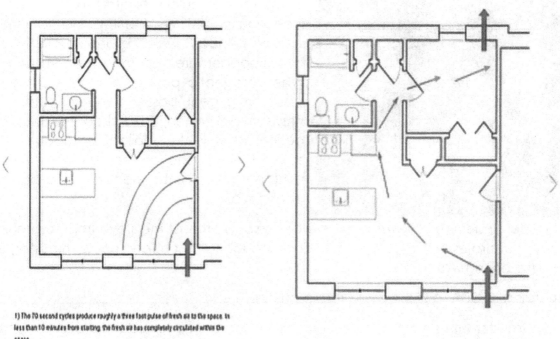

1) The 70 second cycles produce roughly a three foot pulse of fresh air to the space. In less than 10 minutes from starting, the fresh air has completely circulated within the space.

2) The pressure difference causes forced air direction from one side to the other

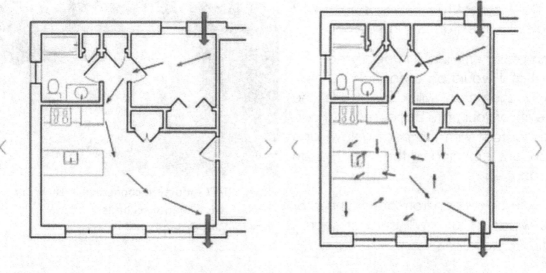

3) And back in the opposite direction.

4) Providing intermediate mixing

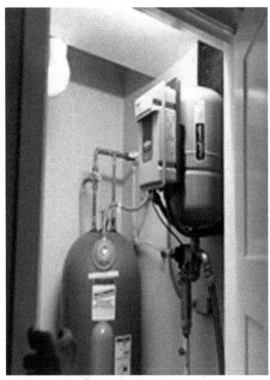

Image 9-16 Marathon hot water heater
Photo: Wes Golomb

Water is heated by this highly insulated and efficient Marathon hot water heater. Like the rest of the appliances and HVAC equipment, it's electric.

John was particularly excited to show me their induction cookstove. The top is a flat, smooth surface beneath which is an induction hob, a coil of copper wire underneath a ceramic plate. When magnetic pots or pans are put on the stove, a current passes through them, which heats them. When I watched this process in action, the water boiled in about twenty seconds. These stoves are extremely efficient, using a fraction of the electricity of a traditional cookstove.

The paper under the pot was put there by Lessa to protect the stove from scratching. It also demonstrates that the stove stays cool enough to touch (and not burn paper) even when it is boiling water.

According to CDA, a British appliance manufacturer:

Energy transfer with induction hobs is around 84 percent compared to around 74 percent for gas or ceramic electric so there are good energy savings. Safety is an important aspect too – there is no naked flame so fire is extremely unlikely.

A pan of water will boil in nearly half the time that it would on a normal gas hob. An induction hob will also ensure the longevity of your pans because they have more contact with the heat below, and the current is running all the way through the pan.

This will stop your pan from developing hot spots which in turn, will burn or scorch food.[15]

Image 9-17 Induction cookstove with water boiling after twenty seconds
Photo: Wes Golomb

**Image 9-18 The kitchen and dining area is sunny by day and lit by three different LED-style lights by night.
Photo: Wes Golomb**

In the above view of the living room, dining room, and screened porch, you can see three different styles of lighting fixtures, each with an LED light. LEDs can use 90% less energy than traditional incandescent light bulbs.

Image 9-19 PV mounted on the side of the garage
Photo: Wes Golomb

Southern exposure would not have worked well on this lot; it was a choice to build the house facing east-west. The original plan was a ground mount PV system set on the knoll to the right of the driveway. John and Lessa later changed their mind on the ground mount. Instead, they changed the slope direction on their garage and put the panels there. The orientation was usable, but the size of the roof limited the area, and thus the generating capacity, of their PV system.

At this time, we have limited information on how much the PV system actually contributes. From first indications, it appears to provide a significant portion of the home's energy. After the system has been running long enough to judge, the owners told me, they might be open to an additional PV array, probably mounted on the ground.

The Finished House

We have talked extensively about the technical details of building the Wallace-Brill house. To the owners of this and other net-zero houses, the technical details provide the basis for their chosen lifestyle, the backdrop of their life.

In this case, that backdrop is a low-maintenance, comfortable, rustic home that accentuates the setting's natural beauty.

Image 9-20 Bedroom
Photo: Wes Golomb

Image 9-21 Alcove and door to screened-in porch; spiral staircase to observatory
Photo: Wes Golomb

Image 9-22 View of the pond from the observatory deck
Photo: Wes Golomb

Almost every building project has challenges, but in the end, all's well that ends well. The owners are extremely happy with their new house. When asked what they would do differently, they could only come up with two minor points. First, Lessa wishes that the windows were a little lower so it would be easier to look at the view while seated. Secondly, she wishes she'd given a bit more thought to the placement of the electric light switches, as they have ended up in some odd places.

**Images 9-23 and 9-24 John and Lessa enjoying the peace and the views from their observatory
Photos: Wes Golomb**

Despite these oversights, which occur to an extent in every house, John and Lessa glowed with joy when they spoke of their new home. It is energy efficient, comfortable, and beautiful.

In speaking with them as well as many other homeowners for this book, I learned that one of the common denominators for a successful project is understanding and paying attention to basic building science principles. And perhaps the most important factor is paying attention to detail. John and Lessa repeatedly told me that working with Bob and his crew was a positive experience. Bob's crew anticipated problems, paid attention to detail, and responded to John and Lessa's questions in a timely fashion. Now that the building process is complete, they live in a house that they love.

Just before this book was published I went back to see John and Lessa. They had lived in the house for almost two years and reported their annual electric bill was around $150/year.

I had assumed because of the small sized pv system, that their bill would be higher. When I inquired, I found that they were augmenting the heat pumps with wood heat.

WALLACE-BRILL

Heated Space: 1900 Sqft
Heat Source(s): Air to Air Heat Pump/Wood

Ratings
Blower Door Ach/Hr: >1
HERS Rating: 31 (prior to solar)
PV: 6 kW

Window Type: Triple Pane
Insulation: Foam/Gasket
U Value: 0.21

Roof
Construction: Timber Frame
Insulation: Rock Wool
R Value: 44

Walls
Construction: Double Wall
Insulation: Cellulose
R Value: 36

F
Foundation: Slab
6" expanded polystrene
R Value: 24

V
Ventilation: Mechanical

$$$
Annual Utility Cost
$150

Built 2019

Image 10-1 The Kliman House
Photo: Wes Golomb

CHAPTER 10

The Kliman Home

I first met Arthur and Debbie Kliman in 2008. After looking at many existing homes, they came to the conclusion that if they wanted a truly energy-efficient home, they would have to build it to their own specifications.

This turned out to be much harder than they expected. The Kliman's reported that most of the builders they spoke with seemed to have no knowledge about building energy-efficient homes and were not interested in this work, apparently because it might take longer and would therefore be less profitable.

After a quite-lengthy process with no luck, Arthur called the local Energy Star office and asked for a referral. He received only two, which, in retrospect, reflects the time. Of the two builders referred, Arthur chose the one he thought would be flexible and genuinely wanted to work with him. Though the Kliman's' design goals were a bit different from the norm, the house was built pretty much as planned.

Like other energy-efficient houses, the Kliman's' house is well insulated, air sealed, and heated efficiently. However, several of these design goals were achieved by different means than those we've discussed so far.

The Kliman's' plan included radiant heat in the floor, which is accomplished by placing PEX tubing, a type of flexible plastic tubing, to carry hot water below the flooring. The hot water heats the foundation slab, which radiates that heat into the house. This makes the proper installation of rigid foam insulation below the floor crucial. When done right, this provides a warm, even heat to the house. When done wrong…

Let me tell you a story. When I was doing energy codes years ago, someone who had just built a similar house came to me complaining that they could not keep their new home warm. Upon investigation, I found that the contractor had neglected to insulate below the slab. Because heat flows from warm to cold and the ground remains around 50°F, most of the heat was flowing from the PEX tubing into the ground. I ended up as a witness in a court case that cost the negligent contractor a lot of money. Make sure to insulate well below the slab!

Images 10-2 and 10-3 ICF ready for installation. The ICF blocks are stacked and locked.
Photos: Arthur Kliman

Insulated Concrete Form (ICF)

While a standard concrete foundation is poured into forms that will be removed, ICFs are both the forms and the insulation for the basement walls. The ICFs stack, and when the concrete has cured, they offer structure, effective air sealing, and continuous insulation with no thermal bridging and an R value of 22.

Image 10-4 ICF with dirt-gravel base
Photo Arthur Kliman

Image 10-5 Insulated floor with PEX radiant heat tubing
Photo: Arthur Kliman

Floor joists are laid in place, and walls and pre-made trusses are installed. Unlike the other homes we've profiled, this house requires structural, load-bearing center beams to support the house above.

Arthur and Debbie used a different strategy for insulation and air sealing than we have seen in the other homes I've profiled. Two inches of spray foam were applied to the walls. The foam provides R-12 of insulation and is an effective air barrier.

The insulating value of the foam is enough to ensure that any moist air that reaches the foam surface will not condense and cause mold, rot, or water damage. After the foam dried, an additional 3.5 inches of fiberglass batts (R-11) were used, providing a total wall-insulating value of R-23 plus the additional minimal R-value of the inside and outside wall surfaces. This combination of materials; spray foam and fiberglass is referred to as flash and batt.

The spray-on foam was also applied to the inside surfaces of the studs, which provides a thermal break. This creates a barrier that stops heat from conducting to the cold outside surface and redirects it into the living space.

Image 10-6 Crossbeam and floor joists in place
Photo: Arthur Kliman

Image 10-7 Walls and prebuilt trusses are installed.
Photo: Arthur Kliman

Spray foam has a high R-value per inch, so it's a very good insulator. It is also very effective at air sealing all the little holes and cracks that are hard to see and reach and can be missed by other methods.

In some cases, such as uneven rock basement spray foam can be the best available alternative. One of the peskiest places to air seal is the sill, spray foam is extremely effective and I've used it for this purpose in my home.

As with most human endeavors, there is also a downside to spray-foam insulation: it is very expensive relative to other insulating materials. This is in part due to the large investment required to purchase the equipment needed to apply the material. Foam has a high Greenhouse Gas (GHG) rating meaning that it uses a lot of fossil fuels in the process of making transporting and applying the material.

Little is actually known about the long-term fate or toxicity of foam insulation. According to an EPA deputy regional administrator, lawyer, and chemist I spoke with, "Foam insulation has never been life-cycle tested, and we do not know if it is the next asbestos."

Image 10-8 Trailer containing spray foam
Photo: Arthur Kliman

Image 10-9 Two different chemicals (isocyanate and polyol resin) are mixed to create foam insulation.
Photo: Arthur Kliman

According to the EPA, "Homeowners who are exposed to isocyanates and other spray foam chemicals in vapors, aerosols, and dust during or after the installation process run the risk of developing asthma, sensitization, lung damage, other respiratory and breathing problems, and skin and eye irritation."[16]

The standard industry guidance, per the EPA, under conditions where the chemicals fully react, appears to be that everyone should stay out of the site for at least 24 hours, and perhaps up to 72 hours, after spray foam is applied. The site should remain well ventilated during this period.

Image 10-10 The foam is applied via tubes.
Photo: Arthur Kliman

Figure 10-11 Two inches of foam applied to a wall.
Photo: Arthur Kliman

As we have discussed, the ideal plan is to define the envelope (the air and thermal barriers along one plane). This means either insulating the floor of the ceiling or including the attic as a heated space and insulating the roof. This was not the case in the Kliman house. The inside of the attic roof was sprayed with an inch of spray foam and the floor of the attic was insulated with cellulose, leaving the air, moisture and a part-thermal barrier on the roof and the bulk of the thermal barrier on the floor.

When you install a heating system, its size depends on the volume of air you need to heat. Arthur's logic was that by insulating the floor of the attic rather than the roof, the volume of air in the attic was excluded from the heat loss calculation and thus a smaller geothermal heat pump system could be installed to heat the living space of the home.

The combination of foam on the roof and fiberglass on the attic floor created a semi-conditioned space between the floor and ceiling of the attic.

My first concern was that with minimal foam insulation in the roof, the inside surface of the foam might get cool enough to allow condensation of water vapor. After many years of inspections, that appears not to be a problem. There is no sign of moisture in the space between the foam-insulated roof and the fiberglass-insulated floor of the attic.

Figure 10-12 Basement above-grade foam-fiberglass walls and foamed sill
Photo: Arthur Kliman

Image 10-13 Below-grade basement SIP walls with two inches of foam applied to the sill.
Photo: Arthur Kliman

The second concern was the lack of an effective air barrier between the floor and attic space. Fiberglass insulation allows air to move through, and that warm air keeps the attic semi-heated. In other words, heated air is leaking out of the conditioned space into the semi-conditioned attic space, causing the heating system to work harder than it needs to.

The Klimans' have not reported an issue with the geothermal system being undersized, but they have reported that they regularly use a gas fireplace for ambiance and warmth. This fossil fuel strategy costs more, emits more carbon and may hide an undersized geothermal system.

Installing the air and thermal barriers on separate planes is not the best practice, and despite the fact that it seems to work in this case, it should be avoided.

Despite these few issues, the combined energy-efficiency measures discussed above serve to keep the energy demands of the Kliman's' house down. They also installed two

Image 10-14 Solar Domestic Hot Water panels on the Kliman home.
Photo: Arthur Kliman

other interesting and effective energy-related features: geothermal heating and solar hot water.

A geothermal heat pump system (as discussed in Chapter 8) takes heat from the earth, concentrates it, and transfers it to a fluid that carries the heat to the house, where that heat energy is stored transferred to and stored in an 80-gallon super-insulated storage tank. This hot water is then used in the radiant floor system to supply heat throughout the house. Interestingly, waste heat from the compressor is captured and transferred to the domestic hot water tank, which is primarily heated by solar energy.

The Klimans also installed a drainback solar hot water system that transfers heat via a flat plate collector to a 66-gallon storage tank. When the tank reaches a temperature of 175°F, the surplus heat is circulated into the larger 80-gallon hot water storage tank that is used for the radiant heating.

The cost of electricity for the Klimans' 2,600-square-foot house is about $3000 a year. That cost is in part due to the Klimans' participation in the Heat Smart program run by their utility. In return for a lower rate to power their geothermal system, the utility has the right to shut off their geothermal heat pump for up to four hours, up to twenty-six times per year. This is a way of limiting demand for electricity at peak demand periods. The only time it has happened has been in the summer. When it does happen, the temperature of the house only rises a degree or so in the four hours the system is off because the envelope is so well air sealed and insulated.

Arthur and Debbie estimate that after federal tax credits and state and utility incentives, the incremental cost of adding energy-efficient and sustainable energy features to their home was about 3.5% of the build cost.

Arthur was quick to explain that the added costs of the energy-related features are more than offset by the savings in operating costs. When I asked what he would suggest to someone wanting to build this type of house, he and Debbie both became quite animated. "The reason we had such trouble finding a builder—and the reason it is still hard to find a builder who builds high-performance homes—is because it takes so much more time and attention to detail." Arthur went on to say, "You cannot be passive about the process. Many builders take the quickest, cheapest way out. You need to specify what you want, and then watch carefully to make sure that you are getting what you specified."

After living in their home for twelve years, Arthur and Debbie are glad they paid attention to details. They remain happy with the comfort of their house, and they love living in the home they designed and built.

In 2008 when the Klimans built their home, Arthur determined that the return on a solar photovoltaic system was not good enough for him to spend the money on it.

They report a utility cost in the range of $3500 a year including their gas fireplace. This is a higher cost than the comparable sized Marion home, primarily because they purchase power rather than generate it.

The Kliman's home is a great example of a Net-Zero ready home! It has a really tight, well insulated shell, All that is needed to make it net-zero is to replace the fossil fuels with solar photovoltaics.

KLIMAN

Heated Space: 2650 Sqft
Heat Source(s): Geothermal/Gas

Foundation: Slab
6" explanded polystyrene
R Value: 24
Basement Walls:
5" Insulated Concrete Forms
R Value: 22

Annual Utility Cost
$3000 Electricity
$500 Gas Fireplace

Ventilation: Mechanical

Walls
Construction: 2x6 Construction
Insulation: 2" Foam, 3.5" Fiberglass
R Value: 23

Roof
Construction: Manufactured Truss
Insulation: Spray Foam/Cellulose
R Value:

Window Type: Triple Pane
Insulation: Foam
U Value: 0.23

Ratings
Blower Door Ach/Hr: 0.6
HERS Rating: 50
PV: No Solar (net zero ready)

Built 2008

Image 11-1 The Burnses' straw-bale home
Photo: Andrea Burns

CHAPTER 11

The Burns Straw Bale Home

I remember well the first time I met Andrea and Jeff Burns in October 2007. I was taking some students on the Green Building Tour. When we arrived at the Burnses' house, they were genuinely excited to see us and proud to show off the house that they built from local materials.

The Burnses built their nearly net-zero home before most people had even conceived of the idea. Built from local materials, heated with wood and passive solar, and powered by 1.2 kW of photovoltaic energy, their off-the-grid home provides sufficient heat and hot water and a good portion of their food.

From the road, the only indication of what's at the end of their driveway is the farm stand where they sell their produce in the summer.

Image 11-2 The Burnses' farm stand
Photo: Wes Golomb

Image 11-3 Driveway to the house, which is barely visible to the left of the barn in the background
Photo: Wes Golomb

Andrea and Jeff Burns live deliberately. They value connection with the land and wanted to build a home with local, non-toxic materials. They desire to live as sustainably as possible and choose which technologies to incorporate into their home. They planned the house to be off the grid, but out of curiosity they got a price of $5000 to run poles and lines from the grid. For the same price they installed a pv system. For the Burnes's then, this decision paid for itself immediately. They have since moved the system from their roof to a ground mount. They have added seven more panels over the past 20 years. Another main reason for staying off the grid was more political.

Jeff, opposed the building of the Seabrook Nuclear Power Plant beginning in 1976, vowed he would not use electricity from, or pay for, the plant. Today, three decades after the Public Service Company of New Hampshire, the utility that built the Seabrook plant, went bankrupt, Jeff and his family live off the grid while the rest of the state continues to pay off this boondoggle every month. When the price of panels came down they added more to charge their batteries faster.

The Burnses' 2,000 square-foot home is built with straw bales, which for centuries were a common building material in the southwestern United States but remain fairly unusual in rural New Hampshire. The Burns family liked the idea of building with straw bales because they are renewable, non-toxic, and have a high R-value. In Alaska, people are building these houses with the bales turned 90° so the walls are even thicker and have a much greater R-value. These houses are also popular in deserts, to black out the sounds of the extremely loud winds.

The main section of the home uses oat straw grown on Tom Abbott's farm just a few miles from the building site, while the addition was built using rye straw grown in Claremont, New Hampshire.

Image 11-4 Roof of house to the right of the driveway's end
Photo: Wes Golomb

Andrea was quick to answer the three most common questions about straw-bale homes.

First, what about fire?

Straw bales are tightly bound and five times more fire retardant than wood. While building the house, the Burnses threw a bale of straw into a bonfire. The next morning, when the fire was out, a singed bale of hay was left in the fire pit.

Second, what about animals?

Straw bales contain no food and are packed so tightly that there is no place for animals to reside.

Third, what about rot or decay?

The bales are coated with plaster, which keeps them dry. The plaster is made from clay, sand, water, and chopped straw to provide strength. The final coat contains lime and sand, which makes it waterproof. The plaster also serves as an effective air barrier and has the wonderful property of wicking any moisture out of the bales.

The result is a well-insulated and air-sealed home built from local materials by the Burns family, friends, and local contractors. In the twenty years that they have lived in the house, the Burnses have experienced none of the above problems.

Image 11-5 First view of the Burns home. In the summer when the sun is high the overhang shades the inside of the house.
Photo: Wes Golomb

Image 11-6 In the late fall, sunlight shining on the tile, heating thermal mass
Photo: Wes Golomb

Passive Solar Design

This home incorporates all of the passive solar design features discussed in Chapter 5, including windows to the south, thermal mass, and an overhang.

In the winter, when the leaves have fallen from the trees and the sun is lower in the sky, sunlight streams deep into the house, warming the air and heating the thermal mass.

Image 11-7 The overhang shades the inside of the house. These photos show that in late fall the sun is low enough to shine into the house to warm the house.

Photo: Wes Golomb

Image 11-8 Living room with mountain view. Note the long window seat in photo 11-8 is filled with sand to create a thermal mass heat storing area.
Photo: Wes Golomb

In the summer, when the sun is high, the overhang shades the inside of the house, As shown in photo 11-5. When the sun goes down, the window treatments are lowered to keep heat in. The well-insulated and air-sealed shell holds the heat, and, as the house slowly cools, the thermal mass radiates its stored heat into the home, keeping it warm. In the summer, when the sun is high, the overhang shades the area that is sunny, as shown in the photos below.

In the next photo (Image 11-9), taken from the outside, you can see that the south-facing side of the house has many windows. When the sun shines in, it strikes and warms the inside floor and walls (thermal mass), which absorb the heat (see Image 11-8 above).

When the sun goes down and the house begins to cool, the thermal mass now radiates the heat back into the room. This is called passive solar heating because the process occurs without any extra input of energy.

Image 11-9 Sun streaming in below overhang on a late fall afternoon. Passive solar design helps to heat the house in winter and shade it in the summer.
Photo: Wes Golomb

Also in Images 11-8 and 11-9, you can see the overhang that shades the summer sun naturally and keeps the house cooler in the summer. As the sun gets lower in the sky, the overhang shades less. By December 21st, the sun will shine directly into the room with no shade, maximizing the heat gain when it is most needed.

The passive solar heating design is augmented by firewood cut from the property.

The wood stove also supplies hot water, which is heated by heat exchangers installed in the woodstove, and then transferred to an insulated storage tank. Hot water from this tank supplies the domestic hot water, and excess is transferred to another, larger tank. Water from this second tank is circulated through tubes under the floor, which radiate heat through the floor and into the house.

Image 11-10 One of two wood stoves that heat the house and the hot water. The copper pipes carry water heat exchangers in the woodstove and back out to the hot water tank.

Photo: Andrea Burns

Unlike the owners of other houses in this book, who chose to use expensive windows, the Burnses elected to use standard double-pane windows. They purchased Warm Window fabric and made insulated window treatments, which you can see rolled up at the top of the windows in the above photo (Image 11-11). When pulled down, these window treatments insulate the windows at least as well as the insulation provided by triple-pane windows. They have the added advantage that they can be lowered in the summer to keep rooms cool.

Andrea reports that people have come into the house and proclaimed, "I didn't know you could air condition an off-the-grid house." In fact, there is no air conditioner. Rather, the combination of insulated windows and a well-insulated, air-sealed shell keeps the house comfortably cool in summer and warm in winter.

Image 11-11 Insulated window quilts can be seen rolled up at the top of the window.
Photo: Wes Golomb

Image 11-12 A sleeping loft in one of the kids' bedrooms
Photo: Wes Golomb

A TRUTH WINDOW

This hole was left in the plaster over the stairs and is the only visible sign that this is a straw bale construction home.

Image 11-13 Stairs to basement with exposed straw
Photo: Wes Golomb

Image 11-14 View from the basement workroom
Photo: Wes Golomb

Image 11-15 On a sunny but cool day in early April the sun is still low enough to send its heat below the overhangs and into the house.

The overhangs also provide outstanding protection for whatever kind of siding you choose to use.

Photo: Andrea Burns

Next page:
Image 11-16 The greenhouse provides food and flowers for the Burns family.
Photo: Wes Golomb

All the electricity for this house is provided by a small photovoltaic system (1.2 kW) and a backup generator, which uses about 100 gallons of gasoline a year (this is the only fossil fuel used by this nearly net-zero home).

Image 11-16 1.2 kW Photovoltaic array supplies all the electricity.
Photo: Wes Golomb

This small input of electricity is a defining factor in the Burns family's lifestyle. You may have noticed that there are few electrical appliances or devices in any of the photographs. The appliances and electronics that characterize most American households are virtually non-existent in this house.

Andrea was clear about it: "I have all the modern conveniences—TV, computer, internet, kitchen appliances, and a sewing machine. I wouldn't want to live any other way. The big difference is that we turn stuff off."

The Burnses have all the modern conveniences they choose and want. However, many in our society would not want to make the same choices. For example, the Burns family cooks all their food on a wood cookstove, and they have no microwave or dishwasher. Many would not elect to live this way, but this is the way of life the Burnses chose. After twenty years of happily living this way, they reaffirm daily that this is what they want.

Jeff showed me their Sun Frost refrigerator and freezer. Both of these appliances run on 12-volt DC current and use about a quarter of the electricity that standard appliances do. To be fair, they are also smaller than most refrigerators and freezers.

I checked in with Jeff and Andrea just before this book was published. They told me that in the last year they did some remodeling of their kitchen, and in the process, pulled apart several of the straw bales. They reported that 21 years after they built the house, the straw was not decomposed or wet and showed no signs of animals. The straw was, as best they could tell, exactly like it had been when they built it.

Net Zero Home

BURNS

Heated Space: 2000 Sqft
Heat Source(s): Wood/Passive Solar

Walls
Construction & Insulation: Post and Beam-Straw Bale
R Value: 36

Foundation:
1/2 slab: polystyrene
(2nd story portion of house)
R Value: 10
1/2 On Piers
(single story portion of house)
Insulated Floor:
Tile - 2-1/2" concrete w/radiant heat
Rough Cut 2x6 boardsm 10" wood shavings
Hardware Cloth, Tyvek

Window Type: Double Pane w/window quilts
Insulation: Straw Bale
U Value: 0.35

Roof
Construction: Standard Construction
Insulation: Cellulose
R Value: 42

Ventilation: Natural (open windwos)

Ratings
Blower Door Ach/Hr: No Test
HERS Rating: None
PV: 1.2 kW

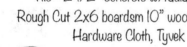

Annual Utility Cost
$150 Gasoline
for backup generator

Image 12-1 The house that Jack built
Photo: Jack Bingham

CHAPTER 12

The House That Jack Built

The Bingham Home, Barrington, NH

Jack Bingham's house is a direct result of the research he conducted at the University of New Hampshire on how heat pumps can make greenhouses usable in the winter. Most greenhouses are either shut down in winter or are inefficiently heated with fossil fuels to keep them open in cold weather. Despite the cold, on a sunny day in February the temperatures in many traditional greenhouses can rise into the high 90s. To cope with this excess heat, the greenhouse is usually vented to keep it cool. In February!

Jack's research centered around installing heat pumps to remove heat from his greenhouse and store it in insulated water tanks. These water tanks were buried in the ground and used when the temperature falls and there was no sun to heat the space. The results were quite amazing. The tanks were able to store 700,000–900,000 British Thermal Units Btu per day, the equivalent of 6–7 gallons of gasoline or 650–750 cubic feet of natural gas! A heat pump that removes energy from hot air is very efficient, and as a result, it can generate this energy for a cost of about 3 cents per kW/hr.

There was one problem, however. Because glass is such a poor insulator, the greenhouses ran out of heat by 3 a.m. That got Jack thinking. Suppose he built a greenhouse attached to a very well-insulated and air sealed home and, using the same concept, captured the heat built up in the greenhouse with a heat pump and stored it in a water tank for future use. And that is what he did! Jack designed a house with a greenhouse on the front, a heat pump, and a storage tank.

But things didn't go exactly as planned. When the builders blasted for a foundation, they found an abundance of boulders.

The boulders made it cost prohibitive to build a basement, so instead they built an Alaskan Slab, also known as a floating slab, in which foam is used as a form under and all the way around the concrete. Then it is covered with a vapor barrier, which is the yellow plastic visible in image 12-4. The black dotted area in the diagram below represents the concrete. The deeper section in the diagram is a 12-inch-deep foundation for the walls, which is poured at the same time and is part of the slab-on-grade. The entire foundation is contained by foam insulation.

Image 12-2 Boulder field below house
Photo: Jack Bingham

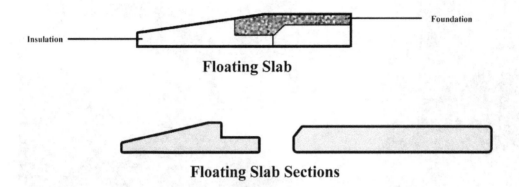

Image 12-3 Floating slab foundation; insulates the slab-on-grade
Diagram courtesy of Jack Bingham

After the concrete cures, the foam remains as an effective insulation barrier, angled down, covered with stone, and extending four feet from the house. This diverts any water away from the house. When complete, the slab is well insulated on each of its exposed surfaces. Reinforced metal bars are then installed for support. The plumbing and drain systems are also installed.

The walls are constructed of structural insulated panels (SIPs), which are made of foam insulation sandwiched between two pieces of oriented-strand board (OSB). They are made in 8-foot-high panels and splined together. They are vacuum-bagged or pressed together to become an impermeable structural component that offers continuous insulation and greatly limits the conductive heat loss experienced in most standard framed houses.

Image 12-4 Alaskan slab. The arrows point to the location of the greenhouse heat-storage tanks.
Photo courtesy of Jack Bingham

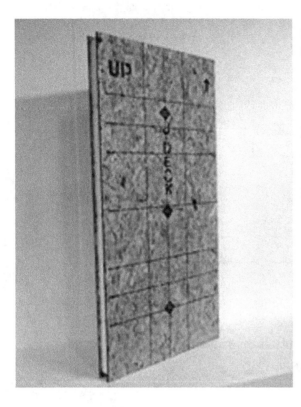

Image 12-5 Structural insulated panel
Photo: Jack Bingham

Image 12-6 Two SIPs splined together
Photo: Jack Bingham.

In standard framed houses, the electric wiring runs through the walls. This often compacts the insulation, making it less effective. Grooves can be cut into the walls to run wires, but it is expensive and challenging to fit the wires through the grooves. To avoid this, the inside of the SIPs are strapped with 2x4's, which allows space for wiring.

The above image shows the house after the panels have been installed. The open area in the center will be the greenhouse. The panels and windows are all pre-cut, built to fit together well and be easily installed. The windows and doors fit in with little or no shimming. It took 8 days to build the house from the slab up to this point.

The 8-inch-thick SIP walls are R-48 and the 12-inch roof is R-72. As

Image 12-7 The house with the panels installed
Photo courtesy of Jack Bingham

mentioned above, there is almost no conductive heat loss through SIPs. Combined with an effective air barrier and a blower door test below 1 Ach/Hr, this is one of the tightest houses I've seen.

You'll recall that in Jack's University of New Hampshire research, even though heat pumps were able to store a lot of heat energy in the water, it wasn't enough to keep the space warm all night due to heat loss through the greenhouse glass on winter nights. The design challenge in this house was to keep that heat in rather than letting it escape. To achieve this, the greenhouse walls, like the rest of the house, were well insulated and air sealed.

At $50,000, double-paned glass was cost prohibitive, so Jack looked into other materials. There are two factors to evaluate when installing glazing—how much light the glazing will let through (its emissivity) and how much heat the glazing will let out (its conduction). These two factors are inverses. With thicker material, the insulation is better but less light gets in. With thinner material, more light gets in but more heat gets out.

The chart below shows the relative options for the greenhouse material. Jack chose the middle option—a three-quarter inch triple-walled polycarbonate.

Production Standards

Thickness (mm)	4	4.5	6	8	10	10	16	20	25	6	8	10	6RDC	20RDC	25	32
Thickness (in)	5/32	3/16	1/4	5/16	3/8	3/8	5/8	3/4	1	1/4	5/16	3/8	5/8	3/4	1	1-1/4
Structure	2 LAYERS				3 LAYERS					4 LAYERS			5 LAYERS		7 LAYERS	
Width (ft)	USA made product available in 47.27", 48", 71.25" & 72" width (Call for special cutting options); Italy made product sold in 82.58" width														47.25"	
Length (ft)					Made in USA Products: Limited to Transportation Requirements											
Weight (lb/ft)	0.564	0.205	0.264	0.307	0.348	0.430	0.582	0.665	0.256	0.317	0.368	0.522	0.635	0.655	0.756	
R-Factor	1.471	1.471	0.264	0.307	0.348	0.430	0.562	0.665	0.256	0.288	0.317	0.356	0.522	0.635	0.455	0.7
U-Factor	0.68	0.68	0.41	0.58	0.52	0.47	0.40	0.38	0.34	0.53	0.48	0.44	0.36	0.33	0.25	0.24
Light transmission %																
Clear	85	84	60	81	82	74	74	75	72	79	79	79	65	65	63	62

Image 12-8 Triple walled polycarbonate options
Image: Jack Bingham

As you can see from the image below, there is a lot of air space contained between three sheets of plastic. The air between the layers of glazing gives the unit the bulk of its insulating value, which is comparable to that of double-paned windows.

Image 12-9 Triple-walled polycarbonate
Photo Jack Bingham

Image 12-10 The greenhouse under construction
Photo courtesy of Jack Bingham

Water Tank for Heat Storage

One of the key features of this design is a means for storing heat after it is collected from the greenhouse. This is accomplished by installing an 800-gallon tank buried under the floor of the greenhouse. The storage tank is contained within a wooden crate made from pressure-treated lumber and shown in the image below (Image 10-11). The tank is insulated to R-28 on the sides and bottom, and it has a cover made from an R-40 SIP. As a result, the hot water stored in the tank holds its heat for a very long time.

Image 12-11 Wooden housing for the water storage tank
Image: Jack Bingham

Image 12-12 Insulated storage tank
Photo courtesy of Jack Bingham

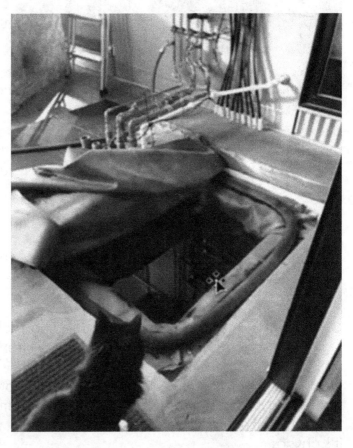

Heat Pump

To recap, the greenhouse is specifically designed to capture the sun's heat and store it for use at a later time. The mechanism used to accomplish this is an air-to-water heat pump installed on the second floor of the greenhouse. Using the same concepts as an air conditioner, the heat pump essentially air conditions the greenhouse. In other words, the heat pump takes heat from the air in the greenhouse and transfers it to water via a heat exchanger. The heated water flows down and into the storage tank, where it deposits the heat before recirculating back up to the heat pump to pick up more heat and repeat the process.

Images 12-13 and 12-14 Air-to-water heat pump mounted near the top of the greenhouse (left). Plumbing black flex tube for radiant heat on the right; copper pipe for heat pump loop. **Photos: Jack Bingham**

discussed the mechanics of heat pumps in detail earlier, but this heat pump has the unique feature of using CO_2 as a refrigerant, rather than using a fluorocarbon as most refrigerators and air conditioners do. As a result, there is no fluorocarbon pollution associated with the process.

A heat pump's efficiency varies with temperature. For years, the challenge of heat pumps has been to make them efficient enough to function in a cold climate. Currently, heat pump technology is effective down to about -10°F or -15°F. Conversely, the higher the temperature, the more efficient the heat pump; fortunately, a sunlit greenhouse is very hot!

On a sunny winter day, it is common to find the greenhouse's temperature at 90°F or higher. At this temperature, heat pumps are extremely efficient. Efficiency is measured by the coefficient of performance (COP), which is the ratio of energy input to energy produced. Jack calculated that for each unit of electricity used by the heat pump, it generated heat equivalent to eight units of electricity! What an incredible ratio: an input of one unit of energy produces an output of eight units of energy! It seems like magic, but it's physics.

The winter of 2019–2020 marked the first winter after Jack's house was built. The system did not come online until well into the cold season. As a result, it took a long time to initially heat up the mass of the house. When it did, the system functioned well. Even with the heat pump working at the high efficiency rates described above, the greenhouse remained hot, although the storage water could have been hotter. In advance of the house's second winter, Jack installed a second heat pump to transfer more heat from the greenhouse to the storage and heating system.

The heat pump's high efficiency makes this system effective and inexpensive to run. To return to an earlier analogy, Jack first sealed the leaky bucket by building a tight, efficient home with efficient devices and then proceeded to keep the bucket filled with a highly efficient heat pump powered completely by solar energy.

As a general rule, switching from combustion of fossil fuels to electricity saves about one-third of the energy that would be consumed by combustion. If grid-tied, the remaining two-thirds will probably be generated from the combustion of fossil fuels (or nuclear power).

This is not an issue at Jack's house, because he has installed 9.5 kW of photovoltaics to generate the majority of his remaining electricity needs. The system is net metered, so when the solar array produces more energy than is needed, the excess is put back onto the grid. This runs the meter backward, so no on-site battery storage is required. With its second heat pump, this house should be net-zero and may possibly be net-positive (generating more electricity than it uses).

When the average person's energy consumption is calculated, a car is the second largest energy consumer after a home. Jack's net-zero home is designed to generate enough energy to also run the family's LEAF electric car.

Jack's house is indeed one of the most unique I have investigated. The synergy created by the tight envelope, extremely efficient heating system, and photovoltaic array provides all the needed electricity to power the house and car. This is an excellent example of the power of these technologies to allow us to live in a comfortable, resilient way without generating pollution.

2022 Update

In the past year Jack added 4.6kW of Solar pv and got the second heat pump working. The system had clearly not been working up to capacity. Since adding the second heat pump, the water in the tank has remained hot, the temperatures in the greenhouse cooler.

One lingering problem remains. Moisture is getting into the air between layers of glazing in the greenhouse. When it is cold outside, the moisture condenses which can cause long term problems and in the short term limits the greenhouse's ability to heat up.

With the added pv installed for part of last year the annual electric bill, including charging the car has dropped to $350 dollars. Jack expects it to drop to 0 in the next full year.

Just prior to publication I spoke with Jack about the house's performance in its second winter. Jack reported that he installed a second heat pump and the system charged quickly and kept the house warm.

They spent about $1000 dollars on utility bills for the second year which included charging their car. Jack is planning on installing another 4.8kW of solar next year with the expectation that they will generate more energy than is needed to run their car too.

Images 12-16 and 12-17 The kitchen with a view into the greenhouse
Photo: Jack Bingham

Image 12-18 Kitchen and dining area
Photo: Jack Bingham

Image 12-19 Living Area
Photo: Jack Bingham

Net Zero Home

BINGHAM

 Heated Space: 1500 Sqft
Heat Source(s): heat pump w/water storage

 Foundation: Slab expanded polystyrene
R Value: 72

 Ventilation: Mechanical

 Annual Utility Cost $1000 (includes charging electric car)

 Walls — **Construction, Insulation:** Structural Insulated Panels
R Value: 48

 Ratings — **Blower Door Ach/Hr:** 1
HERS Rating: -6
PV: 11.43 kW

 Window Type: Triple Pane
Insulation: Foam/Gasket
U Value: 0.20

 Roof — **Construction:** SIP
Insulation: SIP-foam
R Value: 72

Built 2019

Don't Forget!

You can access the free videos that come with this book by going to:

WarmAndCoolHomes.com

CHAPTER 13

Putting It All Together

We've looked at five very different homes. They are made of different materials and have a range of architectural components reflecting their geographical locations and the needs, values, and lifestyles of their homeowners.

When we look carefully, we see that all these homes have quite a bit in common. The most obvious similarity is that they are all owned by people who want to be responsible energy consumers. Each of the homeowners has prioritized approaching or achieving a net-zero energy lifestyle. And, though very different in appearances, the homes all use the same basic concepts:

- Keep moisture away, and out of, the building.
- Define and build a tight envelope that acts as an air barrier, vapor barrier, and thermal barrier.".
- Provide a means for the envelope to dry when it gets wet.
- Provide balanced ventilation to keep the air inside the house clean and healthy. (Most homes have plenty of unbalanced and unregulated ventilation)
- Generate all the energy the house consumes from renewable on-site sources. (Except the Klimans Net-Zero ready home which does not have solar power.)

Looking at the collected data you've seen at the end of each home profile gives us further insight into what is important in net zero homes.

Burns

Net Zero Home
2000

Wallace-Brill

Net Zero Home
2019

Built

Heated Space: 2000 Sqft
Heat Source(s): Wood/Passive Solar

Heated Space: 1900 Sqft
Heat Source(s):
Air to Air Heat Pump/Wood

(F)

Foundation: 1/2 slab: polystyrene
(2nd story portion of house) **R Value:** 10
1/2 On Piers (single story portion of house)
Insulated Floor:
Tile - 2-1/2" concrete w/radiant heat
2" plywood, 10" wood shavings
Hardware Cloth, Tyvek

Foundation: Slab
6" expanded polystrene
R Value: 24

(V)

Ventilation: Natural (open windwos)

Ventilation: Mechanical

Roof

Construction: Standard Construction
Insulation: Cellulose   **R Value:** 42

Construction: Timber Frame
Insulation: Rock Wool
R Value: 44

Walls

Construction & Insulation:
Straw Bale
R Value: 36

Construction: Double Wall
Insulation: Cellulose
R Value: 36

Window Type:
Double Pane w/window quilts
Insulation: Straw Bale
U Value: 0.35

Window Type: Triple Pane
Insulation: Foam/Gasket
U Value: 0.21

Ratings

Blower Door Ach/Hr: No Test
HERS Rating: None
PV: 1.2 kW

Blower Door Ach/Hr: >1
HERS Rating: 31 (prior to solar)
PV: 6 kW

$$$

Annual Utility Cost
$150 Gasoline for backup generator

Annual Utility Cost $150

Marion

Net Zero Home
2017

Heated Space: 2600 Sqft
Heat Source(s):
Air to Air Heat Pump

Foundation: Slab **R Value:** 24
6" explanded polystyrene
Basement Walls: **R Value:** 28
4" polyisosanurate foam

Ventilation: Mechanical

Construction: Manufactured Truss
Insulation: Cellulose
R Value: 62

Construction: Double Wall
Insulation: Cellulose
R Value: 42

Window Type: Triple Pane
Insulation: Foam/Gasket
U Value: 0.19

Blower Door Ach/Hr: 0.6
HERS Rating: 50
PV: No Solar (net zero ready)

Annual Utility Cost $348

Kliman

Net Zero Ready
2006

Heated Space: 2650 Sqft
Heat Source(s): Geothermal/Gas

Foundation: Slab
6" explanded polystyrene **R Value:** 24
Basement Walls:
5" Insulated Concrete Forms
R Value: 22

Ventilation: Mechanical

Construction: Manufactured Truss
Insulation: Spray Foam/Cellulose
R Value: 36

Construction: 2x6 Construction
Insulation: 2" Foam, 3.5" Fiberglass
R Value: 23

Window Type: Triple Pane
Insulation: Foam
U Value: 0.23

Blower Door Ach/Hr: 0.6
HERS Rating: 50
PV: No Solar (net zero ready)

Annual Utility Cost
$3000 Electricity / $500 Gas Fireplace

Bingham

Net Zero Home
2019

Heated Space: 1500 Sqft
Heat Source(s):
heat pump w/water storage

Foundation: Slab
6" explanded polystyrene
R Value: 24
Basement Walls:
5" Insulated Concrete Forms
R Value: 22

Ventilation: Mechanical

Construction: SIP
Insulation: SIP-foam
R Value: 72

Construction & Insulation:
Structural Insulated Panels
R Value: 48

Window Type: Triple Pane
Insulation: Foam/Gasket
U Value: 0.20

Blower Door Ach/Hr: 1
HERS Rating: -6
PV: 11.43 kW

Annual Utility Cost
$1000 (includes charging electric car)

We have data on one house (Kliman) with no solar photovoltaics, and another (Marion) with data before and after solar. Both are within 50 square feet of the same size and we have utility bills for both houses before Mike Marion installed solar.

    Marion 2600 sf Blower door .9 ach/hr Utilities $1500/yr (before solar)
                                                                         $ 348/yr   (after solar)

    Kliman 2650 sf Blower door .6 ach/hr Utilities $3500 /yr

What accounts for these differences in costs? Both houses have very good blower door numbers but the Klimans expenses are significantly higher.

There are several factors that probably effect this data:

The biggest unknown variable is lifestyle. We know the Kliman's use a gas fireplace for both aesthetic and heating purposes, and Debbie likes the house warm.

On the other hand, Mike Marion reports rarely using his heating system even in the winter because of the natural passive solar and the tight envelope.

We don't really know much else about lifestyles

Both homes are well insulated. However, the Kliman's house has no continuous insulation on the above ground walls. As a result, as much as a third more heat conducts out the walls. This does not affect the living conditions, the Kliman's house, insulated and air sealed, stays nice and warm in the winter, but to do so it uses more heat to remain comfy.

Also, an unknown is the cost of the Kliman's geothermal vs. the Marion's air-to air heat pumps to convert electricity to heat. What we do know is that once solar pv was added, Mike Marion's home dropped to just about net zero.

We also see three other smaller homes ranging in design, and size that all are net-zero. Based on this we can make the assumption that If and when the Kliman's decide to add solar pv to their home, they will be able to become net-zero.

Looking at these five houses together I believe the data reveals a few common important factors in building an efficient envelope. They include:

- Moderate to high insulation values including
- Continuous insulation to limit thermal bridging.
- Insulate under and around your cellar floor; insulate your cellar walls.

Good air sealing confirmed with a blower door test, ideally at 1 ach/hr

High quality triple pane windows for added insulation value and increased comfort.

Any home can be made net zero with enough pv on the roof, but by building an efficient envelope one can make the conditioned space comfortable and easier to heat and cool.

The Changing Demographics of Net-Zero Homeowners

A longtime friend of mine has said for years, "There are no net-zero homes, only net-zero people." The Burnses are net-zero people. Twenty years ago, they chose to live off the grid with very low-energy-consuming habits. What you notice immediately when you enter their house is the lack of TVs, stereos, and computers.

When they built their house, living "net-zero" was not only an energy choice but a commitment to a lifestyle. The Burnses' commitment to and love of their lifestyle remains. However, most people would not consider taking up such a lifestyle. Thus, perhaps the most significant change over the last twenty years is that living in a net-zero home no longer requires the same lifestyle choices that the Burnses made. By taking advantage of the decreased costs of solar photovoltaics and using the building techniques described in this book, the option to live in a net-zero home is increasingly attainable. Now, with a sunny location, being a net-zero person is less dependent on a minimalist lifestyle and more dependent upon the number of PV panels installed.

Net-zero homes are ready for prime time. As is the case with Jack Bingham's house, net-zero homes can actually generate more energy than they need, enough to power an electric car or generate excess to be used by others connected to the grid.

With all these benefits, why do net-zero and near net-zero homes remain a small niche market? Why aren't more of these homes being built? I contend that the true and full benefits of living in these homes have not been made clear to most homeowners and housing contractors. There are several reasons for this, and they all come down to a lack of true understanding on the part of key players in the housing market. There is a growing cache of convincing research about the benefits of energy-efficient homes. If understood and assimilated into the housing infrastructure process, this sort of information would clarify many of the benefits we discussed and greatly advance the prospects of net-zero building.

We have considered the benefits to individual homeowners. Let's now turn our attention to the big picture. For the last twenty-plus years, the EPA and a variety of green-building programs both public and private, such as Energy Star, have encouraged the construction of buildings that exceed the requirements of the standard energy code. These homes amount to a small percentage of the total number of homes built over this time, but they provide a large enough sample to compare market values and mortgage risks against standard homes.

Consider this study done in 2019 for the Federal National Mortgage Association[17]:

Using a national random sample, we conducted an analysis of energy-efficient homes rated between 2013 and 2017 and found:

- From the property value analysis, rated [energy-efficient] homes are sold for, on average, 2.7% more than comparable unrated homes
- Better-rated homes are sold for 3-5% more than lesser-rated homes.
- From the loan performance analysis, the default risk of rated homes is not, on average, different from unrated homes, once borrower and underwriting characteristics are considered.
- Loans in the high debt-to-income (DTI) bucket (45% and above) that have ratings, however, appear to have a lower delinquency rate than unrated homes.

In 2013, the University of North Carolina's Center for Community Capital and the Institute for Market Transformation published a paper titled "Home Energy Efficiency and Mortgage Risks."[18] This study examined 71,000 Energy Star and non-Energy Star rated homes and found that default risks were "32 percent lower in energy-efficient homes, controlling for other loan determinants." Among the study's conclusions: "The lower risks associated with energy efficiency should be taken into consideration when underwriting mortgages."

Another study, "Selling Into the Sun: Price Premium Analysis of a Multi-State Dataset of Solar Homes," published in 2015 by the Solar Energy Industries Association, examined:

> "...transactions in eight states that span the years 2002–2013. We find that home buyers are consistently willing to pay PV home premiums across various states, housing and PV markets, and home types; average premiums across the full sample equate to approximately $4/W or $15,000 for an average-sized 3.6-kW PV system."[19]

We already discussed the personal benefits of energy-efficient homes—economics, comfort, health, and resilience. The research above reveals even more benefits! Homes that are energy efficient sell on average for 2.5–5% more than standard homes. Add photovoltaics and it boosts the value of a house by an average of $15,000.

While energy-efficient houses do cost more to build, the operating expenses associated with these houses are lower. Banks can lend to people with a higher debt-to-income (DTI) ratio because of the utility, maintenance, and health expenses that are avoided by living in these houses. So how can we get from building leaky, wasteful houses to building highly energy-efficient homes? Economists call this process market transformation.

MARKET TRANSFORMATION:

Changing the Status Quo on Net-Zero and Near Net-Zero Homes

For hundreds of years, the horse and buggy was the dominant mode of travel. People were employed to shovel horse poop off roads, to build buggies, to feed, maintain, and care for horses, and to dispose of them when they died. In the early 1900s, when humans learned to harness fossil fuels, a transition in transportation began with the introduction of the horseless carriage—aka the automobile. In a very short period of time, society transitioned from the horse and buggy and all of its associated infrastructure to the automobile and a whole new set of infrastructure.

When first introduced, new technologies tend to cost a lot and are adopted only by a few innovators. Early adopters of the automobile saw the advantages of the horseless carriage and followed innovators' lead, buying more cars. As more cars were sold, they were made at a higher quality, more efficiently, and for less money. As a result, the price came down even as the quality increased. With the cost of ownership decreasing and new infrastructure being built, the benefits of cars became more available and affordable to a larger and larger majority of people.

Everett Rogers, a professor of communication studies, popularized this theory about new technologies in a 1962 book called *Diffusion of Innovations*.[20] According to Rogers' theory, a technological transition can be represented by a curve like this:

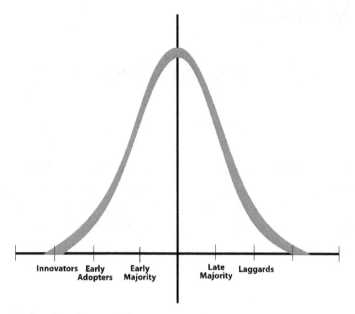

Image 13-1 Technological transition curve

The y-axis represents the number of people adopting the new technology or practice. The x-axis represents time, starting with the first innovators and progressing to the laggards. By the time you get to the right side of the curve, nearly 100% of the population has adopted the change.

Please note that the graph represents new adopters at a particular time. The cumulative number of adopters continues to increase across the length of this graph until we get to the bottom right-hand corner, where nearly 100% adoption has occurred. (Yes, there are still some Quakers or Amish who choose to use the horse and buggy.) The amount of time it takes for a technology to be adopted by a large portion of the population is represented by the length and scale of the y-axis.

According to Malcolm Gladwell, there is a tipping point, which he defines as "the moment of critical mass, the threshold, the boiling point."[21] In the context of market transformation, the tipping point is the point at which the new technology is socially accepted as the norm, the time when many of the last holdouts, as represented by the late majority and laggards on the graph, adopt the technology or practice.

It turns out that if we examine technology adoption patterns for a variety of new technologies, we find a similar curve almost every time. The amount of time over which the transition takes place varies by technology and due to a number of other factors. As you will see, we have a variety of tools at our disposal to help speed the transition.

The transition from horse and buggy to internal combustion engine had a little help along the way in the form of incentives. Starting more than a hundred years ago, policy makers decided that it was in society's best interest to encourage the use of automobiles and thus offered large incentives to oil companies to produce automotive fuels. These oil and gas subsidies continue to this day.

If fossil fuel subsidies were eliminated and the environmental cost of burning fossil fuels was taken into account by, for example, a carbon tax, the true costs of heating with fossil fuels would naturally accelerate the net-zero home market. It would become clear that building to net-zero standards is more economical and a better deal. This, by itself, would be the most effective way to transform the housing market.

As society transformed with the adoption of automobiles, incentives were accompanied by a variety of regulations, including safety regulations, speed limits, licensing, and regulation of a whole new set of infrastructure. There is a strong precedent for the government to both incentivize and regulate renewable energy for the purpose of benefiting society.

One such proposal, an escalating carbon tax which presumably would take away the need for regulation, is being advanced and supported by many economists.[22]

Over the past century, incentives and regulations have often been used as tools that contribute to market transformation. These economic tools have been effective in helping to transform the market for some energy-related devices and technologies. Two examples related to net-zero building are changes in lighting technology and changes in window technology.

When new energy-efficient lighting technology first became available, it was expensive. Through the 1970s and into the 1980s, we used predominantly incandescent and fluorescent lighting. In the 1980s, a transition to compact fluorescent bulbs (CFLs)—which are more energy efficient than incandescents—began and continued through the 1990s. Incentives and regulations sped progress along the adoption curve. Rebates were used to make the cost of new CFL bulbs more affordable. The rebates worked, and within a few years the cost of producing CFLs dropped to a price where the incentives could be reduced and eventually withdrawn. If charted on a graph, the curve would look similar to the one above.

CFLs were a short-lived phenomenon, more recently replaced by light-emitting diode (LED) technology, which through similar incentives offers even higher energy efficiency at an affordable price.

LEDs last many times longer and use approximately 90% less energy than incandescent lights, without the issues of CFLs such as their use of mercury, which needs to be disposed of carefully, a delay when switched on and their inability to dim. Like CFLs, LEDs were very expensive when they first came on the market. Financial subsidies, which brought the price down to an affordable level, were paid by the government and utilities to get people to adopt the new technology.

At the same time, regulations were implemented to raise the standards for minimum lighting efficiency; as a result, there is now limited manufacture of incandescent bulbs. The combination of incentives and regulations accelerated the adoption of energy-efficient lighting.

As the cost of manufacturing LEDs decreases, they become more competitively priced, thereby reducing and ultimately eliminating the need for incentives. Production of less-efficient products diminishes over time, reflecting changes in the market.

People are buying better lighting and reporting their experience of the benefits. Higher demand and greater manufacturing experience are bringing the price down. Regulations limited and then eliminated the most inefficient lighting. Over time, like the transition from horse and buggy to automobile, the lighting transition is progressing.

Changes in window technology tell a similar story. In the last twenty years, window technology has evolved such that triple-glazed windows are not much more expensive than double-glazed windows. You'll recall that windows are measured by their μ-value; the lower the number the better. In 1999, energy codes allowed windows that had a 0.50 μ-value. Today, that has dropped to 0.32 or thereabouts. In 1999, triple-pane windows with μ-values below 0.25 were rare and expensive; today, windows in the 0.18 μ-value range are widely available at a competitive price.

Like automobiles and lighting technology, energy-efficient window technology has improved through a combination of more stringent regulations to limit the lower end of efficiency and incentives to encourage development and use of the best technology.

Each of these market transitions are instructive in how we might approach a transition to net-zero housing. On the adoption curve for net-zero homes, it seems that we are somewhere around the Innovators and Early Adopters phases; in other words, net-zero homes are in the early stages of market transformation. Right now, only a small percentage of home builders truly understand the building science and are willing to put in the extra time and effort required to build tight homes.

There is also a large dichotomy between the general public's perception of net-zero houses and the reality, as seen in the actions of key players such as mainstream lenders, realtors, and appraisers. For example, with so few net-zero homes built, the insurance industry is unprepared to understand and properly value net-zero homes. As a result, the Burns family had terrible trouble finding a company to insure their off-the-grid home. Andrea Burns had the following interaction with a homeowners insurance representative:

> Andrea: *We have solar energy.*
>
> Rep: *You need backup power.*
>
> Andrea: *But our power never goes out. How many people do you know whose power has never gone out?*
>
> Rep: *You need backup power.*

When the Burns family finally found a company that would consider insuring their off-the-grid home, that insurer made them remove the green-living roof. While a green roof is a tried-and-true system used in many places around the world, it was unfamiliar to the insurance company, which ultimately told the Burnses that they have "aesthetic standards to uphold."

The Burns' reported the green roof issue more than ten years ago. One might have hoped things would have changed in that time. Recently, a Vermont insurance carrier turned down a homeowner's policy because, despite thousands of successful installations world wide, the homeowner had installed heat pumps.

Consider my personal experience with a reappraisal of my house. I conducted a deep energy retrofit on my house, thereby reducing the air leakage in my house from around ten Ach/Hr to three. I removed the siding and installed rigid insulation around the walls, which stopped the thermal bridges and doubled the R-value of the walls. I can now heat my house for a fraction of the energy and cost that it formerly required to get through a winter. The indoor humidity stays at a more reasonable level and my sinus issues have abated. In the summer, visitors often think my home is air-conditioned—which it is not—because it stays so cool. I installed 5.6 kW of photovoltaics, which provides me with about 90% of my annual domestic electricity needs. According to the research cited earlier in this chapter, these efficiency retrofits should have raised the house's value between $6,000 and $10,000, and the PV should add another $15,000. Yet the insurance appraiser barely increased the value of my house.

What can be done? Let's begin with specific targeted actions taken by all the players involved in the process of home acquisition.

A Rating System for All Homes

To effectively incentivize and regulate energy-efficient technologies, we first have to measure the technologies so we can compare the effectiveness of our efforts. Windows are measured by their whole-window μ-value, lighting by lumens per watt, and in the case of homes, the standard should be air tightness as measured in air changes per hour (Ach/Hr).

When considering the goal of speeding the market transformation of energy-efficient housing from where it is today to net-zero, the first logical step would be to rate all homes. As discussed in Chapter 1, a variety of rating systems could be used. Most widespread is the HERS numbered rating system, which rates the energy efficiency of a home on a scale from below 0 to 150, where the highest number represents an extremely leaky home. 0 represents net-zero, and the scale extends to negative numbers, which signify a home that produces more energy than it needs, such as a home with enough solar capacity to power an electric car.

Currently, homes with HERS ratings get a sticker, which is typically placed on or near the appliances. If these labels were used on every home and publicized, they would become like a mpg rating on a car—a measure of the projected costs of ownership over a time period and a comparison to other homes. Like labels on appliances, windows, and cars, this would help buyers to understand the value of such homes. Such a rating and labeling system would quickly signal to home buyers the connection between the added costs and benefits of home energy efficiency.

Take the case of two similar-looking houses, the first a standard-code house and the second a net-zero home with a price that is 15% higher. A sticker clearly displayed in each home would reflect the respective house's price and rating. A potential buyer could then weigh the true costs and benefits of the houses under consideration and make a more realistic assessment of the relative value of each home to them.

If information about the efficiency of a home is needed before sale or resale, people must be trained and available to rate these properties. Organizations such as the Building Performance Institute train and certify energy auditors and raters.

Contractors

Understandable, consistent, and enforced energy efficiency standards, regulations, and licensing can help guide contractors, yet to truly transform the market, builders need to be able to see a clear benefit to adopting net-zero construction practices. Benefits might include affordable or subsidized education on net-zero building best practices, as well as incentives for building to a certain energy efficiency standard.

Home Insurance Carriers

As the number of net-zero homes increases, insurance carriers will gather more data and become more comfortable with net-zero home policies. The combination of a rating system that insurance companies can follow and reliable third-party studies related to durability, health, and claims could all assist in making insurance more accessible and easier to purchase.

Realtors

Net-zero or near net-zero homes rarely come on the market. Realtors often have little idea of how to present the advantages of such houses. Further, net-zero homes make the value of the bulk of the houses on the market look bad, so there is little incentive for realtors to push the benefits of these homes.

Following what now seems to be an established theme, perhaps the easiest, most effective tool available to realtors would be displaying the HERS rating in all realty listings. Green Realtor classes are taught in some places, and realtors that attend receive continuing education credits toward relicensing. In these classes, realtors learn some of the building science principles discussed in this book, along with financial analysis to better understand the economics of net-zero homes.

Appraisers

There is a tremendous need for the true value of net-zero homes to be reflected in the marketplace. For this to occur, appraisers must understand the value of something—energy efficiency features—that cannot easily be seen. Therefore, it is imperative that appraiser training includes in-depth coverage of best practices and methods for evaluating the value of a home with modern energy features. A universal rating system like HERS would go a long way toward helping appraisers in their evaluation. The Appraisal Institute has developed a Residential Green and Energy Efficient Addendum.[23] Appraisers should know about, understand, and use this form or an equivalent.

Banks

You can walk into your bank today and get a loan for a new car fairly easily. By contrast, the loan process for a home photovoltaics system takes longer and costs more. Yet this is illogical, because the moment you drive a car off the lot its value decreases, whereas solar power usually pays for itself within six to ten years.

Many banks still do not understand that someone purchasing a net-zero home will have significantly fewer expenses over the course of the mortgage. Heating and cooling will cost less and the house will be more resilient, so it will last longer with fewer repairs needed.

Here again, a combination of ratings, education, and independent studies into the real costs and values of net-zero homes could move banking along the market transition curve.

Consumers

Remember those two houses that appeared the same from the outside, except one was energy efficient and cost 15% more than the other? The average home buyer is not trained to think about energy efficiency, the realtor wanting to make the sale does not bring up the long-term economics of energy efficiency, and the value of the home is not reflected by the appraiser.

The 3–15% higher initial investment in a net-zero home yields a tax-free economic return through savings on energy expenses. This investment is often competitive with the stock market in good times. Rebates and tax incentives are additional benefits of installing measures that bring a home to net-zero standards. Further, these houses are comfortable in both summer and winter, they aren't drafty, and because they are well-sealed, they have fewer insect problems. These benefits must be understood by realtors, appraisers, home insurance companies, and perhaps most importantly by consumers.

The actions described above would lead to better building stock and better-educated consumers who see more than just the initial cost of the home purchase.

There is an ever-growing number of resources available to consumers about net-zero homes. You will find some of these in the resource section Warmandcoolhomes.com website.

Act Locally—Think Globally

Building or retrofitting a net-zero home connects local personal action with the mitigation of climate change.

According to a recent study in the Proceedings of the National Academy of Sciences, "roughly 20% of US energy-related greenhouse gas (GHG) emissions stem from heating, cooling, and powering households."[24] The total carbon footprint of all housing is in reality the net result of millions of people's individual decisions and actions about the amount of energy we use and the resultant amount of carbon we emit into the atmosphere. "By making individual and collective decisions that move us toward net-zero housing, we can help to mitigate the worst impacts of climate change.

NOTES

1. https://teamzero.ort/aboutus
2. https://www.google.com/url?sa=t&rct=j&q=&esrc=s&source=web&cd=&ved=2ahUKEwjgpd632cPtAhVLZN8KHZGGA1gQFjAAegQIAhAC&url=https%3A%2F%2Fwww.climatechangenews.com%2F2019%2F06%2F14%2Fcountries-net-zero-climate-goal%2F&usg=AOvVaw2dUuxPYnnsjc6UA68wq3Fa
3. Watch an interview with LEED inventor David Gottfried at https://youtu.be/9bSKO1DLYDQ/
4. Bob Irving who built both the Marion and the Wallace Brill home is a devotee of Passive House and though he does not get his houses certified by PH, he does adhere to many of the PH techniques and standards.
5. Check out the *Warm and Cool Homes* video series for more detailed information on insulation.
6. Between the time of writing this book and its publication. Bob's most recent house, using Henry Blueskin wrap, tested at around .5 Ach/Hr the best of any house he's built to date.
7. https://www.indepth.energy/blog/archives/08-2015
8. Blog - Building Science, Tips, Techniques, Savings - In https://www.indepth.energy/blog/archives/08-2015
9. To my friends and readers below the equator, substitute north for south and accept my apologies for my "Northerncentric" explanation.
10. *The New York Times*, Dec 25, 2017
11. The Economics of the Tesla Powerwall 2, Charlie Furrer, Stanford University, December 5, 2016, http://large.stanford.edu/courses/2016/ph240/furrer1/
12. RevisionEnergy.Com
13. ReVision Energy's Going Solar Guide for New Hampshire in 2021. https://www.revisionenergy.com/locations/go-solar-in-new-hampshire/
14. Ductless Mini-Split Heat Pump Cost Study (RES 28) Final Report Prepared for: The Electric Program Administrators of Massachusetts Part of the Residential Evaluation Program Area Navigant Consulting, Inc., 1375 Walnut Street, Suite 100, Boulder, CO 80302

https://ma-eeac.org/wp-content/uploads/RES28_Assembled_Report_2018-10-05.pdf

15. How Does Induction Cooking Work? | CDA Appliances. https://www.cda.eu/hobs/how-does-induction-cooking-work/

16. Are spray foam fumes toxic?. https://askinglot.com/are-spray-foam-fumes-toxic

17. Energy Efficiency: Value Added to Properties & Loan https://sf.freddiemac.com/content/_assets/resources/pdf/fact-sheet/energy_efficiency_white_paper.pdf

18. "Home Energy Efficiency and Mortgage Risks," UNC Center for Community Capital & Institute for Market Transformation https://www.imt.org/wp-content/uploads/2018/02/IMT_UNC_HomeEEMortgageRisksfinal.pdf

19. "Selling Into the Sun: Price Premium Analysis of a Multi-State Dataset of Solar Homes," Solar Energy Industries Association, https://www.seia.org/research-resources/selling-sun-price-premium-analysis-multi-state-dataset-solar-homes

20. Rogers, Everett M. (1960), *Diffusion of Innovations*, Free Press

21. Gladwell, Malcolm (2000), *The Tipping Point: How Little Things Can Make a Big Difference*, Little Brown.

22. Economists' Statement on Carbon Dividends https://clcouncil.org/economists-statement/
Signed by 3,589 U.S. economists, 4 former Federal Reserve chairs, 28 Nobel Laureate economists, and 15 former chairs of the Council of Economic Advisers.

23. Residential Green and Energy Efficient Addendum https://www.appraisalinstitute.org/assets/1/7/ResidentialGreenandEnergyEfficientAddendum.pdf

24. "The carbon footprint of household energy use in the United States," Proceedings of the National Academy of Sciences of the United States of America, Benjamin Goldstein, Dimitrios Gounaridis, and Joshua P. Newell https://www.pnas.org/content/117/32/19122

ATTENTION: FUNDING AVAILABLE FOR ENERGY PROJECTS

Just as we were going to print, the US Senate passed legislation that provides funds and incentives for many of the strategies described in Warm and Cool Homes!

As more details become available, we will post funding opportunities, on the same page as the videos.

Additionally, all of the techniques and technologies that are spoken of in this book can be applied to retrofits as well as new houses. Retrofits will be eligible for funding too.

Please go to WarmAndCoolHomes.com and register. You'll get access to more than a dozen videos, funding opportunities, our resource page and added information about retrofits.

WarmAndCoolHomes.com

About the Author

Wes Golomb is a long-time clean energy and climate advocate. While in college he helped organize the first Earth Day in Boston. He's taught environmental sciences for 35 years at a local college. In the 1980's he was trained as an energy auditor and since that time has continued to use his interest and knowledge of building science in several capacities including administering the NH Energy Codes office.

Wes has a long record of volunteer service to the State of NH and his Local Community. He has served on a variety of public and non-profit energy/environment boards including 12 years on the NH Sustainable Energy Association's board and the State of NH Energy Efficiency/Sustainable Energy board. He is Chairman of his local energy committee, and 23-year member and past Chairman of his town's Conservation Commission.

In 2006 Wes started the first two-year Energy Services degree Program in the Northeast. The program trained students for careers in Energy Efficiency and Sustainable Energy and better than a 90% graduate placement rate. Wes won awards from the New England Board of Higher Education as well as the NH State Senate for his work.

Since retiring Wes started *TheEnergyGeek.org*, an informational website about energy and climate related matters. His blog, <u>Sustainability Matters</u> has had contributions from writers all over the world. He is available (hybrid) for speaking, workshops and consulting.

In addition to energy work, Wes is an avid photographer and musician. He lives in Deerfield NH with his wife Laurie and his dogs Luke and Buster.

CPSIA information can be obtained
at www.ICGtesting.com
Printed in the USA
JSHW060313260623
43667JS00001B/5